Electronic integrated circuits

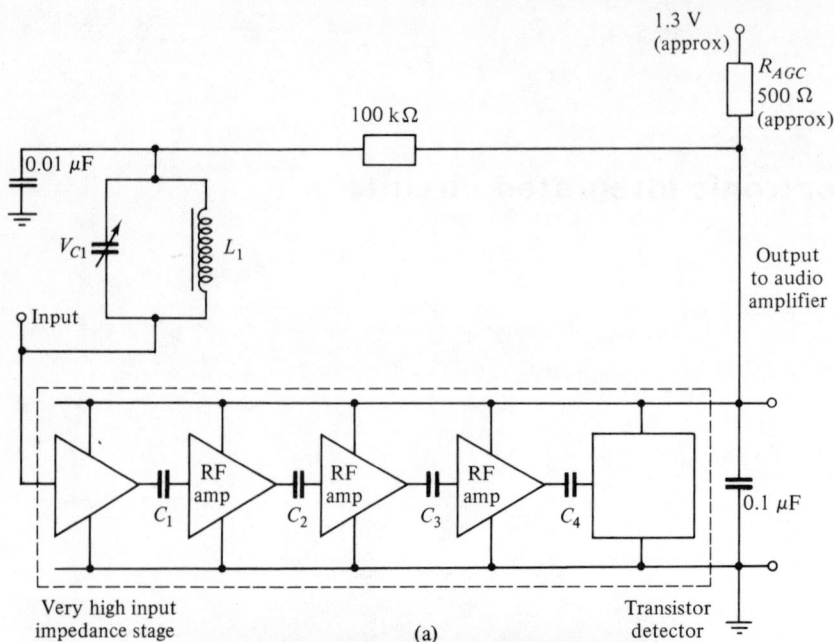

(a)

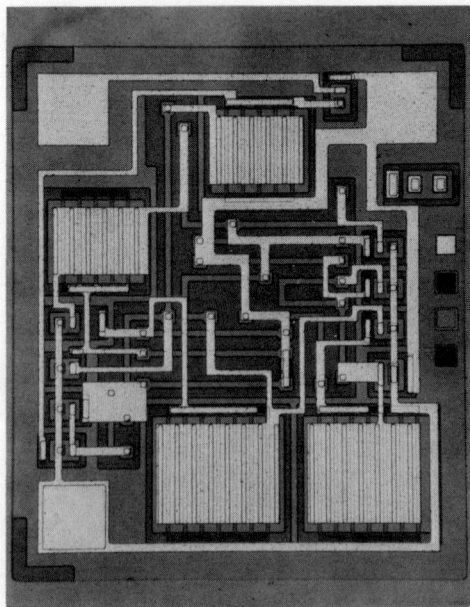

(b)

A CDI integrated radio receiver. (a) A block diagram of the integrated circuit radio. (b) A microphotograph of the integrated circuit chip which has dimensions 0.75 x 0.75 mm and incorporates 30 components, including 10 transistors. Reproduced by kind permission of Ferranti Ltd, Electronic Components Division.

Electronic integrated circuits

Their technology and design

John Allison

Senior Lecturer
Department of Electronic and Electrical Engineering
University of Sheffield

McGraw
Hill

London · New York · St Louis · San Francisco · Düsseldorf
Johannesburg · Kuala Lumpur · Mexico · Montreal · New Delhi
Panama · Paris · São Paulo · Singapore · Sydney · Toronto

Published by McGRAW-HILL Book Company (UK) Limited

MAIDENHEAD · BERKSHIRE · ENGLAND

Library of Congress Cataloging in Publication Data

Allison, John.
    Electronic integrated circuits.

    Bibliography: p. 135
    Includes index.
    1. Integrated circuits. I. Title.
TK7874.A43   621.381'73   74-28367
ISBN 0-07-084051-2

10 9 8 7 6 5 4 3 2 1

PRINTED AND BOUND IN GREAT BRITAIN

Contents

Preface

Electronic engineering is a relatively young science and has always been subject
to rapid changes in the technology and function of electronic devices. Never in its
history have these changes been so marked as in the last decade, since the introduction of integrated circuits. During this time, electronic engineers of all disciplines have become involved with the technology and design of integrated circuits
and subsystems. While there is nothing unusual in their having to learn about and
incorporate novel components, what is perhaps exceptional is that the circuit
designer, as well as the manufacturer, is now involved at some or all technological
levels of circuit production.

The systems engineer no longer designs using discrete components, but employs complete prefabricated integrated subsystems. At the very least, he must
be aware of the advantages and limitations of the various types of integrated
circuit and their component parts. These properties are intimately connected with
the manufacturing method. The subsystem circuit designer has to become involved
in the technology of integrated circuit production even more deeply and at an
earlier stage, since many integrated circuits are now made which incorporate some
degree of customer design at some point in their manufacture. Inevitably, the
designer of a basic integrated circuit is intimately and inseparably concerned with
the fine detail of the technology of integrated circuit manufacture.

The aim of this book is to provide a simple, readable introduction to the design
and fabrication processes of integrated circuits. It is based on a series of lectures
given to final year undergraduate students. Much of the necessary background on
the properties of electronic devices and materials is presented to them elsewhere.*
However, the practising electronic engineer will already be aware of sufficient
elementary semiconductor physics and properties of transistors to enable him to
read this book without much difficulty. The book serves the twofold purpose,
therefore, of initiating students of electronic engineering and related physical
sciences into integrated circuits, and also providing practising engineers with an
introduction to the subject.

The book first describes the various types of integration technique, paying
particular attention to the silicon planar, diffused, integrated circuit process,
which has now become firmly established and is likely to dominate for many
years. Whereas there are bound to be some advances in the details of the

* See, for example, Allison, J. *Electronic Engineering Materials and Devices*, McGraw-Hill
  1972.

technology, the basic processes of slice preparation, diffusion, epitaxy, window preparation and so on, are always likely to figure in any development. This is the reason for including a fairly detailed account of these techniques in the central sections of the book. Next follows a description of the circuit components that are available in the various technological systems, together with a discussion of their properties and important design parameters. Methods for electrical isolation between integrated circuit components and packaging are then described and compared. A brief account of the recent derived technologies and components is also included, taking, as examples, the collector-diffused isolation, ion-implantation and silicon-gate processes and the charge-coupled device. Finally, a section on integrated circuit design is presented. It is prevented from being fully comprehensive by limitations of space, but serves as a simple introduction to the subject, emphasizing the inevitable influence that a particular technology has on circuit design, and vice versa.

I am indebted to my colleagues, in particular Mr R. W. J. Barker, for many helpful discussions.

I would like also to express my appreciation to Mrs E. Byrne, Mrs B. Cowell, Mrs M. J. Knowles, Mrs D. Loukes, and Miss H. J. Taylor for their unstinting help with the preparation of the manuscript and to thank Monica, my wife, for invaluable editorial assistance and endless cups of black tea.

Finally, may I venture to hope that readers of this book will find it as interesting and informative as I have found it enjoyable to write.

John Allison
May 1974

1. Integrated circuits

1.1 Introduction

Electronic engineering is a relatively new science, beginning essentially with the invention of the triode in 1906, but changing dramatically with the discovery of the transistor just over 25 years ago, and more recently, with the development of field-effect devices, integrated circuits, and the latest solid-state components. Because of this short time scale and the continual introduction of new devices and techniques, electronics has always been a rapidly varying field, requiring constant updating of education and experience. Perhaps that is its fascination; it is always modern, up to date, 'with it', while retaining as its basis sound, unchanging physical fundamentals.

At no time has this rapid advance been more apparent than during the last decade, since the introduction and universal acceptance of integrated circuits. This latest phase has been described as revolutionary; in view of the complete re-appraisal that has been necessary by electronics engineers of all disciplines, whether they are concerned with the technology of circuit production, circuit design, or systems applications, and because of the impact made on the consumer market, the description is not inappropriate.

Why has the introduction of integrated circuits brought about such a startling electronics revolution? The answer lies in the overwhelming advantages to be gained by circuit integration. What are these advantages? Lightness, compactness, ability to function from low-voltage battery supplies are obvious considerations. This answer, conditioned by the appealing application of integrated circuits to problems of space-travel and computers, is perfectly correct, and indeed such projects would be difficult to envisage, if not impossible, without integrated circuits, but it is not complete.

Another, possibly more important, consideration is the usual dominant engineering criterion, economics. Since many thousands of components and their interconnections which form numerous individual circuits, or alternatively a sub-system, can be fabricated in one small slice of semiconductor by a series of well-developed, inexpensive steps, to be described in detail later, the cost of complete circuits or even subsystems is potentially very small.

A further not unimportant advantage of integrated circuits is their inherent reliability and long life. These attributes stem initially from the increased reliability gained when transistors superseded thermionic valves which, in spite of extensive development, suffer from inherent weaknesses, due to fragile electrodes and encapsulation, and short cathode and filament lifetimes. The obvious superiority of transistors over valves or electro-mechanically operated switches, such as relays, is exploited in integrated circuits. An additional advantage is that all component interconnections, which for discrete circuits are most prone to failure, for example, at soldered joints, are fabricated simultaneously, to become an integrated part of a complete circuit.

Another advantage is that it is possible for an integrated circuit to achieve performance levels, for example in high speed logic, which would be inconceivable and often unrealizable with discrete circuits, and to perform circuit functions, for example charge-coupled delay (described later), which would be impossible by conventional circuit techniques.

Although there have been profound changes in the semiconductor industry since its inception, both in materials and techniques, the broad basis of technology has now been established, and silicon, planar diffused integrated circuits are likely to dominate electronic engineering for many years to come.

At present, integrated circuits occupy about one-third of the semiconductor market but it has been estimated that when world sales of semiconductors rise to £3000 M by 1980, two-thirds of this amount will be expended on integrated circuits.

Even at their present level of complexity, integrated circuits are much more important than other electronic components, because of their ability to incorporate a circuit subsystem in one small package, which in turn can control the performance and economics of a total system. The system designer must therefore be conversant with manufacturing technology, since his design will be critically dependent on the capabilities of integrated circuits. Further, since custom-designed circuits, which today comprise 10 per cent of the U.K. integrated circuit market for example, are expected to expand to 50 per cent of the market in the mid-seventies, together with a corresponding growth in the customer-designed integrated circuit market, close liaison between user and manufacturer will be essential. Again, the circuit designer must be aware of the capabilities and limitations involved in the technological processes of integrated circuit manufacture.

1.2 Types of integrated circuit

Integrated circuits may be classified according to the manufacturing processes employed for their fabrication and to their circuit function. General descriptions in the former category include *thick- and thin-film, monolithic, hybrid,* and *multi-chip* integrated circuits. Other descriptive adjectives are often appended to specify the technology and materials of particular circuits in more detail, for example, planar, diffused, junction isolated, silicon monolithic integrated circuits. Additionally, a circuit description, for example, an array of transistor–transistor logic NAND gates, completes the specification.

2

1.2.1 Film circuits

In film circuits, all components are fabricated by depositing passive or active layers on to a single common supporting insulating slice, or *substrate*, conducting interconnections being made by a superposed metallic film. In their *thin-film* version, the substrate is often a ceramic or glass slice, e.g., alumina, and the layers are deposited by vacuum evaporation, r.f. sputtering, electroplating or other thin-film techniques. Alternatively the various layers which constitute each component and its interconnections can be deposited on to an insulating substrate by printing techniques, using, for example, a screen process, in which the inks used are so formulated that they reduce to resistive or highly conducting films, etc., after a subsequent heat treatment stage. This method of fabrication is called the *thick-film* process.

The major difficulty which frustrated early attempts to produce complete all-film integrated circuits was the lack of suitable active film devices which were compatible with the rest of the technology. Whereas thin-film transistors, for example, are now available, complete thin-film integrated circuits do not appear to be becoming generally accepted, nor do they have the potential for large-scale integration that is inherent in monolithic circuits, so further discussion of film techniques and components will be restricted to their use to complement other types of circuit, for example monolithic structures and hybrid or multichip integrated circuits.

1.2.2 Monolithic integrated circuits

Monolithic (Greek: single-stone) circuits are so called because all the active and passive components are formed as an integral part of a single semiconducting substrate, usually silicon. The active layers in the silicon which constitute each component are produced by modifying the material by compensation doping successive layers with impurities, the dopant being introduced by diffusion or epitaxial techniques, to produce, for example, n–p–n bipolar transistor structures and derivative components in the silicon. Interconnections are again made by depositing an interconnecting pattern of a conductor such as aluminium, using thin-film techniques, or alternatively they can be formed from highly doped, low resistivity paths in the bulk parent semiconductor itself.

Because each circuit and its component parts can be made so extremely small, many identical circuits can be fabricated simultaneously on one slice of silicon, or alternatively many different types of circuit can be made on a slice and interconnected with a metallizing layer to produce a complete electronic subsystem. An idea of the sizes involved can be obtained from Fig. 1.1. Each silicon slice of, say, 50 mm diameter can accommodate about 1800 integrated circuits, each occupying typically a 1 mm square chip of semiconductor. A bipolar transistor, which may be a component of a particular circuit, occupies an area of say 0.15 x 0.1 mm and other components can be smaller, so around fifty circuit elements per individual circuit are possible. If the slice diameter is increased to, say,

3

100 mm, which is the current trend, then the number of components and circuits per slice can be considerably increased. In addition, if a technology such as that used for metal-oxide-silicon (MOS) circuits, or alternatively an improved bipolar process such as collector-diffused isolation (CDI), is employed, then the area of devices can be reduced and the packing density of components and circuits increased correspondingly.

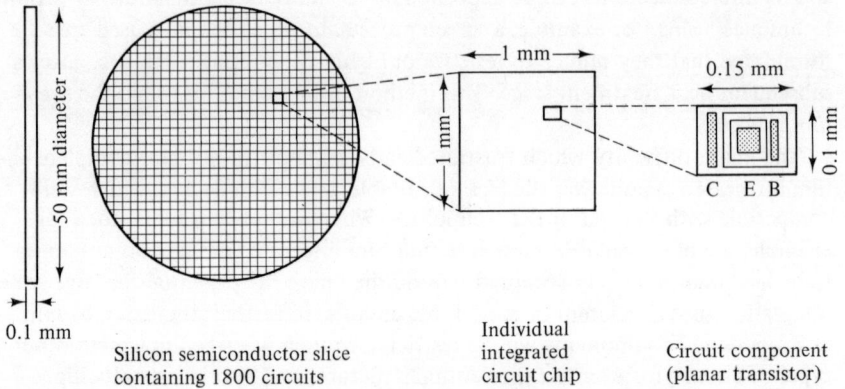

Silicon semiconductor slice containing 1800 circuits

Individual integrated circuit chip

Circuit component (planar transistor)

Fig. 1.1   Typical geometry of silicon slice, integrated circuits, and components.

Since all circuit components are constructed by compatible planar diffusion processes and are all formed at the same time by a sequence of photolithographic and metallurgical steps, circuit reproducibility, reliability, and economy are inherent features of the monolithic technology.

1.2.3 Hybrid and Multichip Integrated Circuits

Hybrid integrated circuits use thin or thick film techniques to form passive components and interconnections on an insulating substrate but the active components, such as transistors and diodes, are produced separately, for example, in silicon by a planar process, the unencapsulated active semiconductor chips being designed for direct attachment to the interconnections of the thin-film circuit.

In an alternative technology, the *multiple-chip* integrated circuit consists of monolithic or thin-film circuits, or parts of circuits, which are bonded to a common substrate or header and interconnected, say by bonded gold wires, to produce a complete circuit or subsystem.

Again, since the monolithic, silicon, planar, diffused integrated circuit is now established as the principal technology, any further discussion of hybrid and multichip techniques will be limited to those areas in which they support or complement the dominant monolithic process.

4

1.3 Monolithic integrated circuits

As discussed earlier, the principal technology now established for the production of integrated circuits is that which employs silicon, monolithic, planar integrated processes. A brief outline of the basic processing steps used for the fabrication of a monolithic circuit will be given in this section and later chapters will be devoted to discussing these processes in more detail.

The fabrication of the simple hypothetical circuit shown in Fig. 1.2(a) will be used to illustrate the technology. The various processing stages which are to be described are shown sequentially in the cross-sectional drawings of Fig. 1.2, which should be studied in conjunction with the flow chart shown in Fig. 1.3. For clarity, the particular circuit chosen has only a few components, oriented in such a way that interconnections can be seen clearly on the sectional drawings; in a more general circuit there would be many more circuit components and the interconnections might lie in any direction on the surface of the semiconducting slice.

The starting material for monolithic integrated circuits is a flat polished slice of p-type semiconductor, usually silicon but sometimes germanium, or exceptionally, gallium arsenide. Typically, the silicon slice might have a resistivity of 0.05 Ω-m with dimensions 50 mm diameter and thickness 0.1 mm, Figs. 1.2(b) and 1.3(b). In common with most solid-state devices, it is essential that this material is pure and single-crystal, so as to avoid masking the fundamental semiconductor action by otherwise dominant spurious effects associated with non-crystallinity, unknown impurities, metallurgical defects, and so on.

A thin layer of n-type silicon, typically 0.005 Ω-m, 10 μm thick and doped with arsenic donors of concentration about 10^{22} m^{-3}, is subsequently grown on one face of the p-type slice, in such a way that it is both single crystal and also its crystal structure is a continuation of that of the parent material, Fig. 1.2(f). This process is known as *epitaxy* and an *epitaxial-* (or sometimes *epi-*) layer results. Many individual circuits and all their component parts will eventually be fabricated in this epitaxial layer. The p-type slice, known as a *substrate*, not only serves to support mechanically the thin epi-layer, but, as we shall see later, also assists in electrically isolating from each other circuit components built into the layer.

A feature of planar integrated circuit technology is that all connections to devices and circuit interconnections are made in one plane on the face of the epi-layer. This could result in long high-resistance paths, for example to the active collector regions of planar transistors, which would degrade circuit performance. Highly conducting n^+ layers (the + sign indicating relatively heavy doping levels) are diffused under all active devices to help to overcome this difficulty. These so-called *buried layers* are diffused into the p-type substrate before the epitaxial n-type layer is grown, Figs. 1.2(d) and (e) and 1.3(d) and (e). Doping concentrations of typically 10^{25} donors m^{-3} produce the highly conducting layers required.

The buried layer diffusion illustrates two processes that are basic to the whole technology. First, the n^+-type layer is produced by doping what was previously a

5

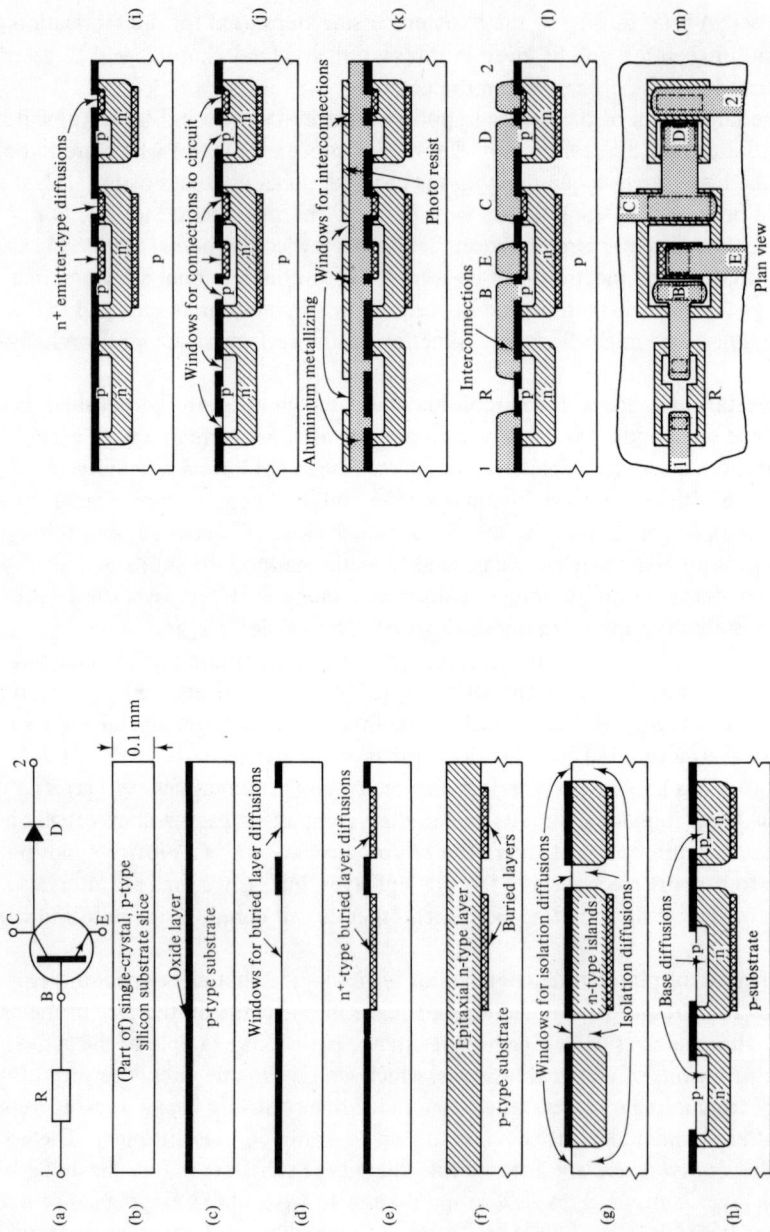

Fig. 1.2 Fabrication of a monolithic integrated circuit.

6

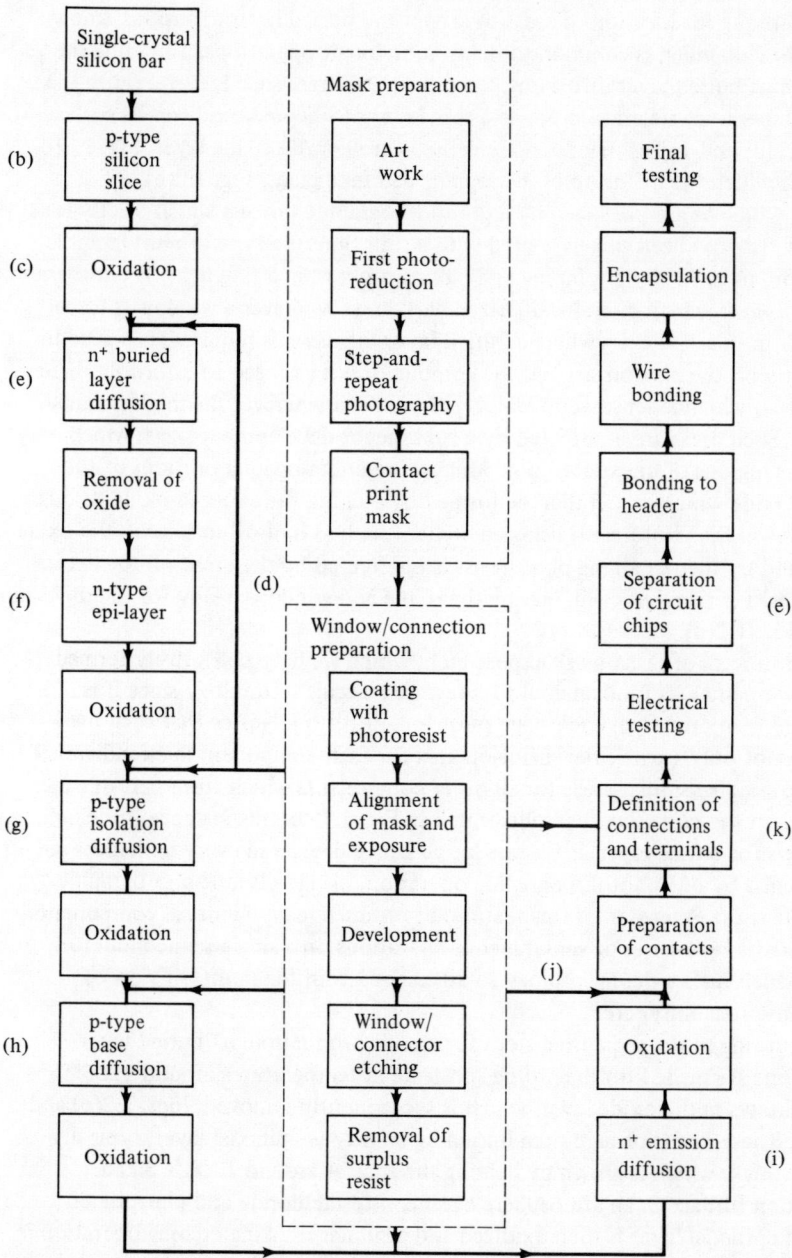

Fig. 1.3 *Flow chart of some of the basic steps for the production of monolithic, planar, diffused, silicon integrated circuits.*

p-type semiconducting substrate with sufficient donor impurities to cause a reversal of its conductivity type, a process known as *compensation doping*. Second, the technique for location of areas to accept the buried layer diffusions, known as *window opening*, is common to many subsequent operations. To define the location of buried layer diffusions, the p-type substrate slice is first oxidized in high-temperature steam to produce a thin layer of silicon dioxide on its surface, Figs. 1.2(c) and 1.3(c). Windows are etched through this oxide layer, Fig. 1.2(d), through which the diffusion of n^+ dopants can take place, Fig. 1.2(e).

The windows are produced by a photolithographic process which briefly is as follows. The oxidized slice is coated with a thin film of a type of photographic emulsion, or *resist*, by a spinning and baking process, in a safe light. A photographic plate or *mask*, which contains life-sized dark areas wherever a window is to be opened, in this case everywhere a buried layer diffusion is required, is placed in contact with the photoresist and the combination is exposed to ultraviolet light. The resist, which is sensitive to UV, is polymerized wherever the mask is transparent. Such areas are unaffected by a subsequent development stage, which only removes regions of unexposed film which were under opaque portions of the mask. Oxide windows can then be formed by etching the exposed oxide through the holes in the photoresist, using an etchant such as hydrofluoric acid; the oxide protected by the remaining photoresist is unaffected by the etch. All the excess photoresist is then removed, leaving the oxide and windows ready for the diffusion stage, Fig. 1.2(d).

This process of resist-mask-expose-etch, which we have collectively termed window opening, is fundamental to integrated circuit technology since it is repeated many times, at each stage prior to a diffusion. Notice that each mask defines not only a particular diffusion area for each component in an individual circuit, but also similar areas for all other components which form part of other circuits on the same slice, as is illustrated in Fig. 1.1. In this instance, for example, windows for buried layer diffusions for all active devices in every individual circuit are opened by one photolithographic operation, using only one photographic mask. It is the extension of this feature, to produce many identical components and circuits by a series of simultaneous operations on a single semiconductor slice, which makes the technology so attractive, from the points of view of economy, reliability, etc.

Returning to the sequential steps for circuit production, n^+ buried layer diffusions are made into the p-type substrate at temperatures around $1200°C$, via windows in the oxide layer, which is subsequently removed, Figs. 1.2(e) and 1.3(e). These diffusions are then buried by an n-type epitaxial layer, typically $10 \ \mu m$ thick, which is grown by heating the slice at around $1200°C$ in an induction furnace in an atmosphere of silicon tetrachloride and phosphine.

The epitaxial layer is then oxidized and a similar masking-etching operation is used to cut windows through which a p-type isolating diffusion can take place, Figs. 1.2(g) and 1.3(g). Of course a completely different photographic mask must be employed, but it must be made and used in such a way that there is good

8

registration between it and the previous masking stage. The isolation diffusion is normally a two-stage operation in which boron dopant is firstly deposited at high temperature on the surface, the slice then being oxidized and the dopant *driven-in* at around $1000°C$, to compensation dope regions of the epi-layer to change it to p-type. The diffusion is arranged, by controlling temperature and time, to penetrate the epi-layer completely and join the substrate. Islands of n-type epitaxial material surrounded by electrically isolating p-type, are produced in this way; it is in these islands that all circuit components are eventually fabricated.

Then follows a now familiar masking-window-cutting operation to define areas in the n-type islands for p-type diffusions to produce transistor bases and monolithic resistors, Figs. 1.2(h) and 1.3(h). This diffusion is typically about 2 μm deep, and boron is again the usual acceptor dopant, this time in a lower concentration.

Next, oxidation, window cutting and an n^+ diffusion using phosphorus dopant creates all emitters, collector contact areas and similar component parts, Figs. 1.2(i) and 1.3(i).

All components for the monolithic circuits have now been fabricated and interconnections between them have next to be made. The interconnections are usually formed in aluminium which is again shaped by a photolithographic process. The semiconductor slice is first oxidized and then windows are opened to define connections to the various component parts in each circuit, e.g., emitters, bases, resistors, and so on. Aluminium is then vacuum evaporated over the entire slice, say to a thickness of 2 μm, making contact to the circuit components via the oxide windows. The aluminium is then coated with resist, exposed via a suitable mask and developed, Fig. 1.2(k). Surplus aluminium is then etched away through the windows of the photoresist, using, for example, a caustic soda solution, to define the circuit interconnection pattern and aluminium terminal pads for connection of individual circuits to external leads, Fig. 1.2(l) and (m).

Each circuit or subsystem is usually tested electrically at this stage, while still on the slice. This is because processing costs per circuit have so far been relatively small, so deficient circuits are best eliminated before the next, costly, packaging operation. A test head containing many needle probes, each connected to measuring instruments or power supplies, is lowered on to the aluminium terminal pads of each circuit in turn, and automatic measurements are made of its major electrical characteristics. For complex circuits, this testing operation has to be computer controlled, and sometimes statistical sample testing has to be resorted to because of the many circuits involved. Any circuit which is found to be not within specification is marked on the slice and eventually discarded.

The next stage is the separation of each individual circuit on the slice from its neighbour (unless the circuits are to be interconnected on the slice to form an integrated subsystem, as described later). Separation is achieved by diamond scribing and cleaving to produce small *chips* or *dice*, each containing a complete integrated circuit, Figs. 1.2(l) and (m).

Finally, each circuit chip is fixed to a suitable header and fine wires, of gold or aluminium, are bonded to the circuit terminal pads and header lead-throughs, to form connections between the monolithic integrated circuit and the outside world. The final product is encapsulated, for example by encasing the header and circuit in a thermo-setting plastic. This is followed by circuit testing and quality assessment.

Subsequent chapters will describe in more detail each of the basic operations used for circuit fabrication.

2. Preparation of semiconductor slices

The preparation of the basic semiconductor slices, in which all integrated circuit components are fabricated in subsequent operations, is described in this chapter. The production of pure, single-crystal semiconductor, impurity doping, epitaxial layer growth and evaluation, all of which are stages in the fabrication of the basic starting material from which planar monolithic integrated circuits are manufactured, will be discussed in turn.

2.1 Growth and refining of silicon crystals

2.1.1 Raw materials

Since most monolithic integrated circuits are now fabricated in silicon, our discussion will be limited to the preparation of substrates made from this semiconductor. The most commonly occurring natural sources of silicon are its oxides, e.g., sand and quartz. These can be reduced in a furnace with carbon to produce 98 per cent pure silicon. However, this impurity level is much too high for semiconductor device preparation; a typical requirement might be for less than one impurity atom per 10^9 silicon atoms for the starting material, so the commercial silicon requires considerable refining before it is suitable.

2.1.2 Zone-refining

One technique used for the purification of industrial silicon is known as zone-refining. The apparatus used is shown schematically in Fig. 2.1. The silicon is cast into a long, thin ingot which is selectively heated locally, usually by an induction heating coil, to produce a short molten section. The molten silicon is prevented from separating from the ingot by surface-tension forces. Since most impurity atoms in silicon have an affinity for the liquid rather than the solid state, they are trapped in the narrow molten zone; hence, if the heating coil is slowly traversed along the rod, taking the molten region with it, the impurities are confined to this region and are swept to one end of the bar. After several such passes, unwanted impurities are almost entirely concentrated at the ends of the bar, which can be subsequently cut off and discarded.

11

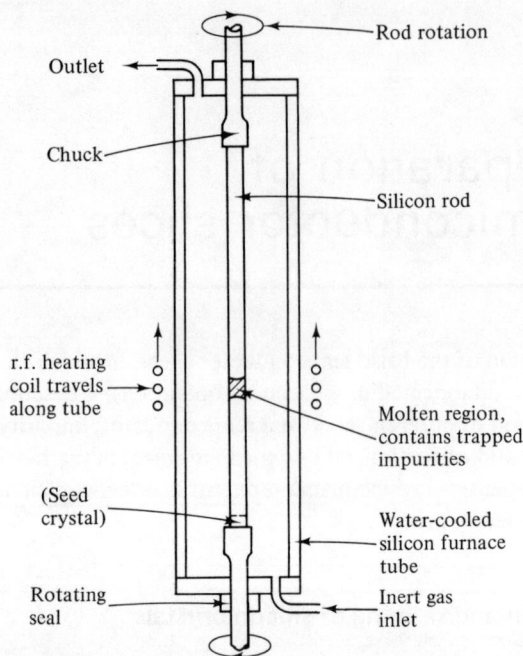

Fig. 2.1   Schematic diagram of zone-refining equipment.

Features of the zone-refining process are that no crucibles are required and that the entire operation is carried out in an inert atmosphere, both of which tend to prevent additional contamination by impurities.

2.1.3 Crystal growth from the melt

Although the silicon is at this stage highly refined and free from impurities it is still polycrystalline, and a necessary requirement for substrates is that it is single-crystal semiconductor. Single-crystal silicon can be prepared by controlled freezing from a melt. A difficulty is that the concentration of atoms in the molten material is very much greater than in the regular diamond lattice of the crystalline form; this precludes methods of crystal growth in crucibles, which would result in a material containing many dislocations.

A procedure for growing silicon crystals which has found wide commercial use, the *Czochralski* method, obviates this difficulty. A correctly oriented seed crystal is partly immersed in molten refined silicon. The melt temperature is then reduced slightly until silicon begins to freeze on the cooler seed crystal, which is then slowly withdrawn, see Fig. 2.2. If the temperature and withdrawal rate are correctly chosen, the liquid–solid interface remains near to the surface of the melt and a long single-crystal of silicon is pulled from it. This process is also carried out in an inert atmosphere, probably argon or helium, to prevent oxidation.

12

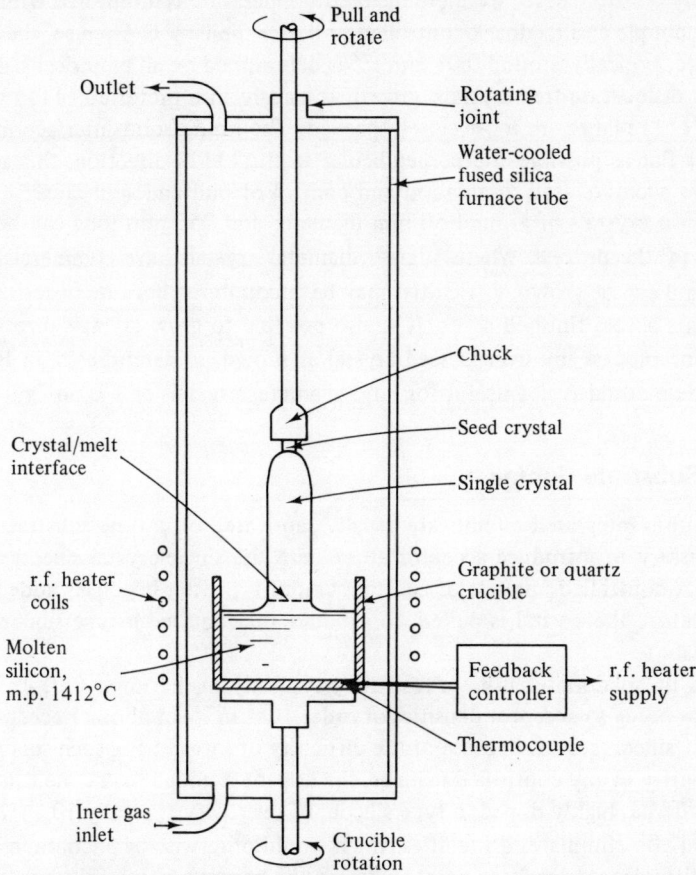

Fig. 2.2 *Schematic diagram of apparatus for the preparation of single-crystal silicon by the Czochralski method.*

A further refinement is that both melt and puller are continuously rotated, to produce a more homogeneous crystal.

If H_i is the heat input to the system per second, H_l the heat lost per second and a length dx crystallizes in time dt, then the heat balance equation is

$$(H_i - H_l)\,dt = L\,.\,d_s\,.\,A\,dx$$

where L is the latent heat and d_s the density of silicon and A is the cross-sectional area of the crystal. Rearranging gives the pull-rate,

$$\frac{dx}{dt} = \frac{H_i - H_l}{L d_s A}$$

Hence we see that the pull-rate is closely related to the heat input and losses, crystal properties and dimensions. The conditions for crystal pulling are therefore carefully controlled, for example the melt temperature is monitored with a thermocouple and feedback controlled to better than $\pm\frac{1}{2}°$C. Even so, the final pull-rate, typically around 0.01 mm s^{-1} is determined on an empirical basis.

For dislocation-free crystals, growth is usually in a preferred $\langle 111 \rangle$ direction,* since (111) planes are most closely packed. The liquid/solid interface must be kept as flat as possible and perpendicular to the $\langle 111 \rangle$ direction; this again requires accurate seed orientation and control of pull and spin rates.

Silicon crystals of around 50 mm diameter and 250 mm long can be produced by the process. Whereas larger diameter crystals have commercial advantages and can be grown, difficulties may be encountered because of resistivity gradients across finished slices. It is also possible to grow crystals directly in the zone process, by using a seed crystal as shown, in parentheses, in Fig. 2.1, but this method is not useful for large diameter crystals and is not much used.

2.2 Substrate doping

Monolithic integrated circuits are usually fabricated on p-type substrates, so it is necessary to introduce acceptor atoms into the single-crystal silicon at some stage. A controlled amount of acceptor impurity, often boron, is added to the melt before the crystal is pulled, to produce the required p-type silicon substrate material.

The finished material has a resistivity typically in the range 0.01–0.5 Ω-m, corresponding to acceptor densities of order 10^{21} m^{-3}, or about 1 acceptor atom per 10^7 silicon atoms. Because of the difficulty of introducing such small quantities of impurity in any controlled manner, the dopant is added to the melt in the form of powdered, highly doped, p-type silicon, of resistivity typically 10^{-4} Ω-m. This method also eliminates difficulties which would otherwise be encountered due to large differences in melting points between the parent and impurity materials.

A further consideration, which makes the growth of uniformly doped crystals of prescribed resistivity difficult, is that the concentration of dopant in the solid phase of the crystal is much greater than in the liquid phase. In other words, as a silicon crystal is pulled, the melt becomes richer with time.

The conductivity of the doped semiconductor material is given by the usual expression:

$$\sigma \simeq ep\mu_h = eN_a\mu_h \tag{2.1}$$

which assumes a hole mobility of μ_h, a hole density, p, an acceptor concentration of N_a, and that all the acceptors are ionized at room temperature. However, the mobility is to some extent a function of acceptor concentration, so the evaluation of the conductivity is largely empirical, using design curves of the type shown in Fig. 2.3.

* For further details of crystallographic notation see Anderson, J. C. and Leaver, K. D., *Materials Science*, Nelson, or most other books on this subject.

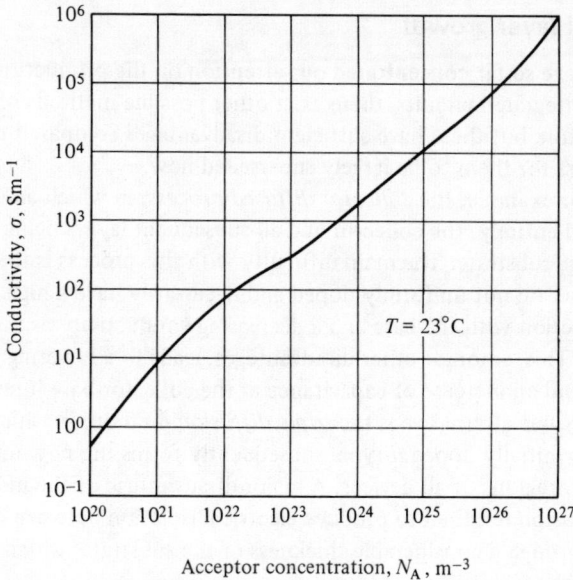

Fig. 2.3 Example of a design curve for estimating the conductivity of p-type silicon from a knowledge of the impurity concentration.

Because of the various empirical growth parameters mentioned, prediction of the resistivity of the final substrate material can only be done to an accuracy of around ±25 per cent.

2.3 Substrate slicing and polishing

Substrate slices, typically 0.1 mm thick, are cut with a diamond saw from the single-crystal silicon bar. Correct orientation of the surface of the slices with respect to the crystal planes is important, particularly for successful epitaxial layer growth at the next stage, so slices are cut as close to the (111) or sometimes the (100) plane as possible. These slices which, as cut, are quite rough, are then lapped to remove saw marks and to produce a flat surface. Surface damage, which after this operation can still exist to a depth of around 20 μm, is removed with a chemical etch employing an acid mixture consisting of nitric acid to oxidize the surface and hydrofluoric acid to dissolve the oxide.

The slices are then polished mechanically on a wheel to a mirror-like finish, using alumina abrasive powders of decreasing grit size down to a final 1 μm diameter. Surface damage, which is still present down to around 2 μm deep, is finally removed by an additional chemical etching stage, which can sometimes be simultaneous with the final polishing stage.

The finished slices are usually checked for resistivity at this stage, using a four-point probe method which is described later.

15

2.4 Epitaxial layer growth

Although we have so far concentrated our attention on the production of epi-taxial, planar integrated circuits, there exist other possible methods of monolithic circuit fabrication, but these have sufficient disadvantages compared to the epitaxial process for them to be largely superseded now.

Consider, for example, the *collector diffused process*, in which an epitaxial layer is omitted entirely, the collector and all subsequent layers being diffused directly into the substrate. The main difficulty with this process is that the collector regions are not uniformly doped and necessarily have a high conductivity near to the junction with the base and a decreasing conductivity as the substrate is approached. This, amongst other disadvantages, leads to a lowering of the break-down voltage and an increase of capacitance at the collector-base junction.

Another possible alternative is the *triple diffusion* method, in which the sub-strate, which is initially doped n-type, subsequently forms the now uniformly doped collector regions of all devices. A major disadvantage of the method is that the p-type diffusion required to produce electrical isolation between components must diffuse through a considerable thickness of the substrate, which takes a relatively long time.

The *epitaxial planar* process, described in chapter 1, obviates the difficulties encountered in both of these alternatives and provides components with improved parameters and circuits with a better performance. A p-type silicon substrate supports a single crystal, n-type epitaxial layer, which provides uniformly doped collector regions with no limitations on thickness and whose electrical properties can be independently controlled by the doping levels in substrate and epilayer, so as to optimize the characteristics of incorporated devices. The techniques used to form and evaluate such epilayers will be described in the following sections.

2.4.1 Epitaxy

Epitaxy, a Greek word meaning 'arranging upon', is a process for the growth of a thin, single-crystal layer of doped semiconductor on to a substrate. Deposited atoms arrange themselves along existing planes of the crystalline substrate material, bonding to parent atoms to form an unbroken extension of the crystal structure. The structure of the grown epitaxial layer is thus a continuation of that of the single-crystal substrate.

In the epitaxial process, a thin layer of silicon in which all circuit components will be formed eventually, is grown on a thicker substrate slice which provides mechanical support. Not only can the epi-layer have a different doping concentra-tion and hence conductivity than the substrate, but the conductivity type can be different, while still retaining the orderly crystal structure. For example, the epi-taxial layer is n-type, with typical resistivity 0.005 Ω-m, whereas the substrate is p-type with resistivity around 0.1 Ω-m. Choice of doping concentration in the epi-layer, which determines its resistivity, is arranged so as to optimize the per-formance of transistors formed in it, since all collectors are made from this material. The designed resistivity is an engineering compromise between the requirement for a high value, and hence high breakdown voltage of built-in transistors, and a low value which would reduce the collector access resistance, and hence improve the high frequency performance and lower the collector saturation voltage, $V_{CE(sat)}$, which is the voltage existing between collector and emitter when the transistor is driven hard on into saturation.

Because of their different conductivity types a p–n junction is formed at the interface between epi-layer and substrate, which, as we shall see, becomes part of the electrical isolation system separating components in the layer. Typical dimensions and resistivity ranges are shown in Fig. 2.4.

A further advantage of the process is that a doped epitaxial layer can have a much more uniform impurity concentration than a diffused layer. Although it is

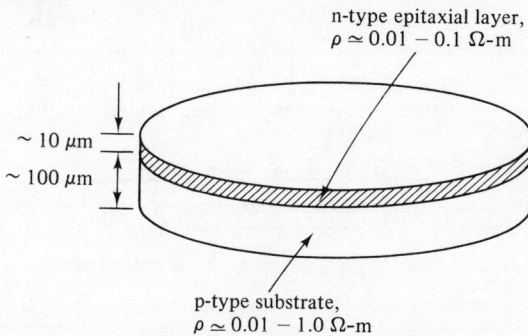

n-type epitaxial layer,
$\rho \simeq 0.01 - 0.1\ \Omega\text{-m}$

$\sim 10\ \mu m$

$\sim 100\ \mu m$

p-type substrate,
$\rho \simeq 0.01 - 1.0\ \Omega\text{-m}$

Fig. 2.4   An epitaxial layer on a silicon substrate slice.

possible to grow successive epi-layers to exploit this advantage, this is accompanied by some deterioration in crystallinity, so most monolithic integrated circuits have only one epilayer into which successive layers are grown by diffusion techniques.

There is no additional difficulty in growing epi-layers over diffused areas, so that, for example, growth over the n^+ buried layers previously diffused into a substrate presents no problem.

2.4.2 Epitaxial growth of silicon

Although silicon atoms can be deposited directly by vacuum deposition or sputtering techniques, the most widely used technique utilizes the vapour-phase reduction of silicon tetrachloride. Silicon atoms are transported by one of these methods on to a substrate, which is held at a high temperature to enhance their attachment and so speed up the process.

In the vapour-phase technique, silicon chloride vapour is reduced with hydrogen on the surface of silicon substrate slices, which are heated in an r.f. induction furnace, to produce silicon indirectly:

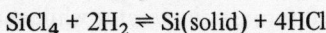

$$SiCl_4 + 2H_2 \rightleftharpoons Si(solid) + 4HCl$$

A schematic diagram of the apparatus used is shown in Fig. 2.5. The temperature of the reactor is kept, typically, at 1200°C. If the temperature is reduced much below this value, the epi-layers become more defective and eventually non-crystalline, because of the reduced mobility of the deposited atoms and their subsequent difficulty in moving to correct sites in the crystal lattice.

Note that the chemical reaction indicated by the equation is reversible. Hence, if hydrogen chloride vapour is present in the carrier gas, etching rather than epitaxial growth occurs. When the concentration of silicon tetrachloride is high, etching can still occur even when hydrogen chloride is not present, due to a competing interaction:

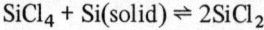

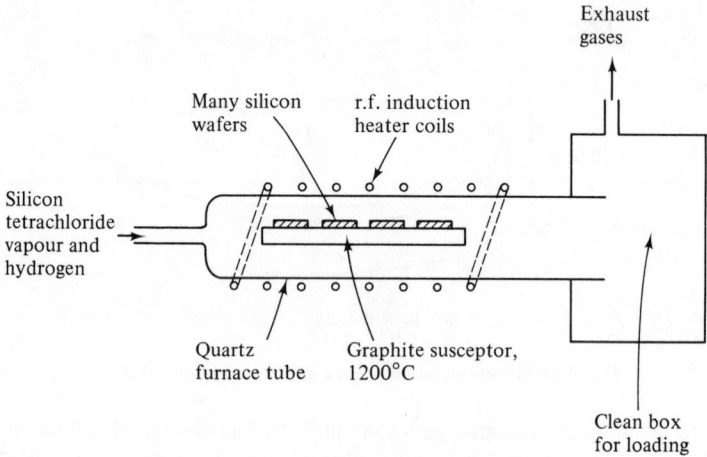

$$SiCl_4 + Si(solid) \rightleftharpoons 2SiCl_2$$

Fig. 2.5 *Reactor for epitaxial silicon layer growth.*

Thus the growth rate of epitaxial silicon, which will be negative if etching occurs, is very critically dependent on the concentration of silicon chloride as well as the temperature, as shown in Fig. 2.6. It will be noted that the typical industrial conditions for growth at a rate of around 1 μm per minute produce layers which are well within the region for single-crystal epitaxy.

Doping of the epitaxial layer is achieved by adding controlled amounts of the appropriate impurity in liquid form, for example phosphorus trichloride or arsenic trichloride, to the silicon chloride, or the dopant can be introduced directly into the reactor chamber in gaseous form, for example as phosphine gas, PH_3.

The additional equipment for epi-layer growth and doping is shown in Fig. 2.7. A typical production schedule might be briefly as follows. Substrate slices are cleaned by etching in a hydrofluoric–nitric–acetic-acid mix, followed by washing in deionized water and spin drying in nitrogen. Such a slice cleaning operation is usually included between successive stages in the production of integrated circuits. An inert gas such as argon or nitrogen is passed through the reactor while the

18

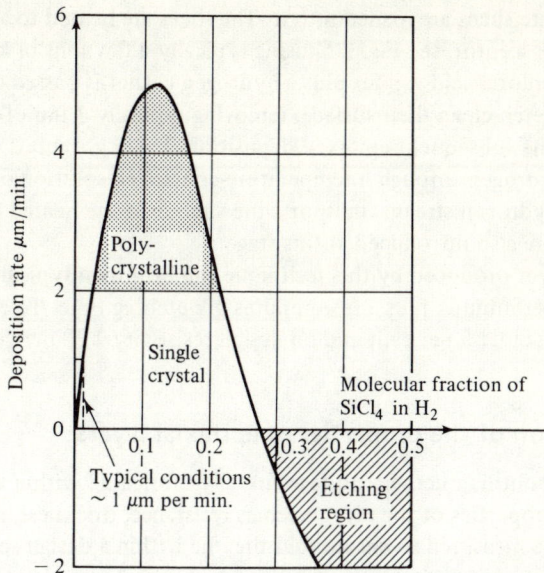

Fig. 2.6 *The variation of growth/etching rate with concentration of silicon chloride, at some particular temperature.*

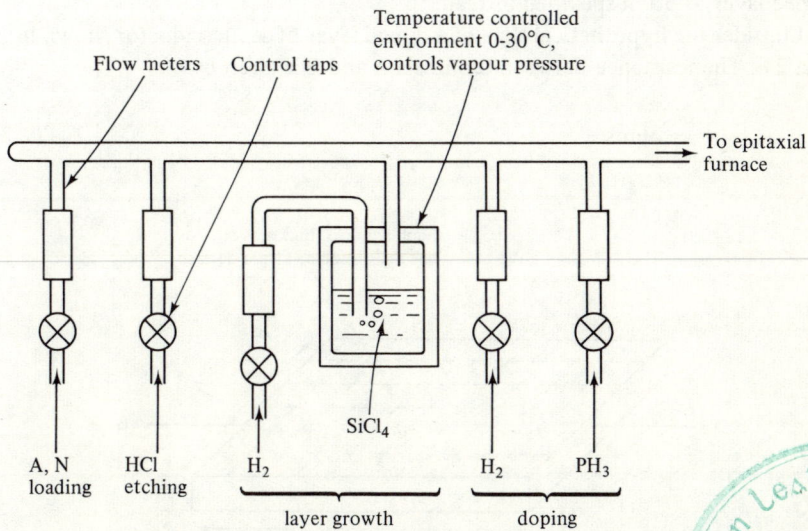

Fig. 2.7 *Auxilliary supplies for an epitaxial reactor.*

19

cleaned substrate slices are loaded into it. The slices are heated to around 1200°C via a graphite susceptor, see Fig. 2.5, using typically a few tens of kilowatts of r.f. power. Hydrochloric acid vapour plus a hydrogen carrier is passed over the substrate slices to etch-clean their surface, removing typically 1 μm of material and preparing for the subsequent epitaxial deposition. Layer growth is then achieved by bubbling hydrogen through a temperature-controlled solution of silicon chloride; the hydrogen stream transports the vapour to the heated substrates. Dopant gases are also introduced at this stage.

A 10 μm layer produced by this technique can be grown typically at a rate of around 1 μm per minute, plus a few minutes pre-etching time. Tolerances on the thickness are about ±5 per cent and on design resistivity ±20 per cent.

2.5 Evaluation of the properties of epitaxial layers

The entire monolithic microcircuit is eventually fabricated within the epi-layer so the electrical properties of the layer, such as resistance, thickness, resistivity profile, must be measured to ensure that they lie within a design specification. The measurement concepts and techniques which are used and are later described are also generally applicable to the characterization of diffused layers.

2.5.1 Sheet resistance

Because epitaxial or diffused layers are not necessarily homogeneously doped, it is not always appropriate to quote their conductivity, which is usually a function of distance through the layer. A *sheet resistance*, ρ_s, corresponding to a particular doped layer, is often specified instead.

Consider the hypothetical slice of a doped layer of semiconductor shown in Fig. 2.8. The resistance between terminals A and B is given by

$$R_{AB} = \frac{\bar{\rho}l}{wt} \text{ ohms} \tag{2.2}$$

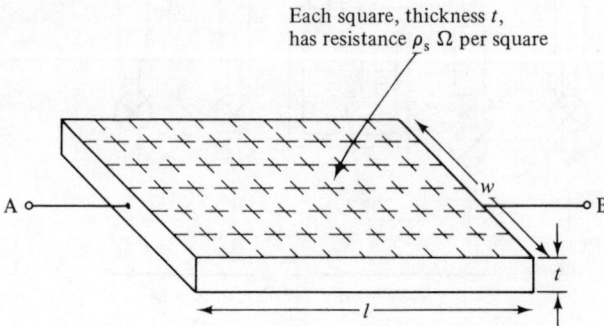

Fig. 2.8   Portion of an epitaxial or diffused layer, thickness **t.**

20

where $\bar{\rho}$ Ωm is the mean resistivity of the layer. Alternatively the resistance can be found by dividing the surface into squares, each of thickness t, and defining the resistance of each square as the *sheet resistance*, ρ_s Ω/square. The value of ρ_s can be found by putting $w = l$ in eq. (2.2), which gives

$$\rho_s = \frac{\bar{\rho}}{t} \tag{2.3}$$

This is a constant which is independent of square size, i.e., the squares can be 1 mm x 1 mm, 1 μm x 1 μm, etc. Since the entire slice consists of columns of l squares in series of resistance $\rho_s l$, and there are w such columns in parallel, the total resistance between A and B is then:

$$R_{AB} = \frac{\rho_s l}{w} \text{ ohms} \tag{2.4}$$

The advantage of specifying resistance in this manner is that the resistance of a doped layer of particular dimensions can be determined without a knowledge of the layer thickness or conductivity profile, provided the sheet resistance is known. This technique is particularly useful for the design of monolithic resistors, as we shall see later.

2.5.2 Measurement of sheet resistance

The sheet resistance of epitaxial layers is most conveniently measured using a four-point probe method. A schematic diagram of the apparatus used is shown in Fig. 2.9. A fixed, measured current, typically 1 mA, is passed between the two

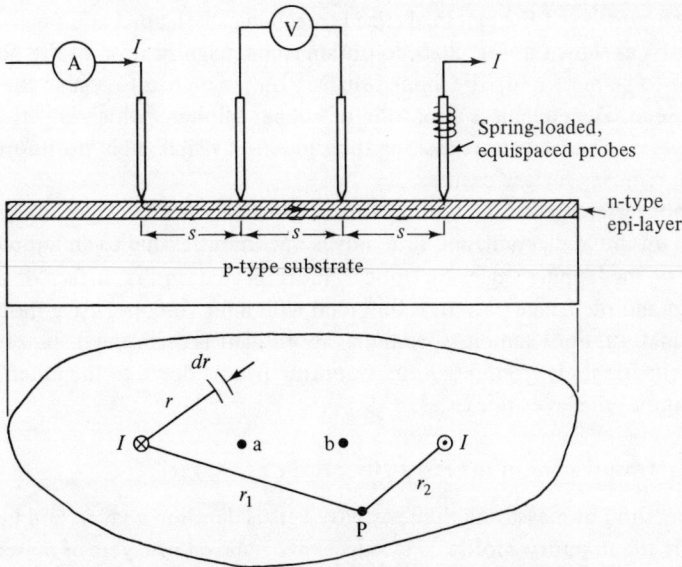

Fig. 2.9   *Four-point probe measurement of sheet resistance.*

outer probes and the voltage, V, is measured, using a high input impedance voltmeter, between the two inner probes. Current flow is restricted to the epitaxial layer because of the opposite conductivity type of the substrate.

The resistance of the elemental annular ring shown in Fig. 2.9 is $\rho_s \, dr/2\pi r$, so the volt drop across the ring is:

$$\mathrm{d}V_r = I \frac{\rho_s \, dr}{2\pi r} \tag{2.5}$$

The potential at the arbitrary point P, obtained by integration of this expression is:

$$V_p = \frac{I\rho_s}{2\pi} \log_e \left[\frac{r_2}{r_1} \right] + \text{constant}$$

so the potential difference measured, V, is given by:

$$V = V_a - V_b = \frac{I\rho_s}{\pi} \log_e (2)$$

which can be rearranged to give:

$$\rho_s = \left[\frac{\pi}{\log_e 2} \right] \frac{V}{I} = 4.532 \frac{V}{I} \tag{2.6}$$

Hence the sheet resistance can be estimated from a measurement of V and I.

2.5.3 Measurement of epitaxial layer thickness

A small portion of the epitaxial slice is mechanically lapped at a small angle, of order $1°$, as shown in Fig. 2.10, to obtain some magnification of the boundary region to be measured. The lapped surface is then etched to enable the junction to be seen. The etchant is a solution of copper sulphate, which selectively plates the epi-layer and substrate making their junction visible, in hydrofluoric acid which removes surface oxides.

An interferometric method is often used to measure the layer thickness directly, as shown in the diagram, and so removes uncertainties due to an imprecise knowledge of the lapping angle. An optical flat is located on the surface of the slice as shown and the fringe pattern is observed with a microscope, using monochromatic illumination. For example, if sodium vapour light is used, the distance between adjacent fringes is around 0.3 μm. Counting fringes down to the junction thus determines the layer thickness.

2.5.4 Measurement of the resistivity profile of a layer

One method of measuring the resistivity versus depth in a layer, and hence determining the impurity profile, is by successive removal of layers of material, measuring the new sheet resistance after each step, possibly by a four-point probe method.

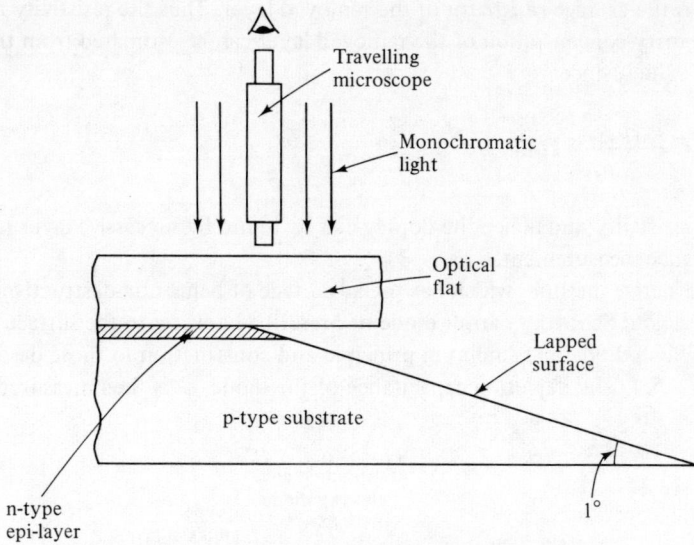

Fig. 2.10   Optical measurement of epi-layer thickness.

Referring to Fig. 2.11, the initial resistance between the ends of the sheet is:

$$R_{AB1} = \rho_{s1} \frac{l}{w}$$

but after removal of a layer of thickness δ the resistance changes to:

$$R_{AB2} = \rho_{s2} \frac{l}{w}$$

The resistance of the removed layer, R_{ab1}, is then:

$$R_{ab1} = \frac{R_{AB1} \cdot R_{AB2}}{R_{AB2} - R_{AB1}} = \frac{l}{w} \frac{\rho_{s1}\rho_{s2}}{\rho_{s2} - \rho_{s1}} = \frac{\overline{\rho_1} l}{w\delta}$$

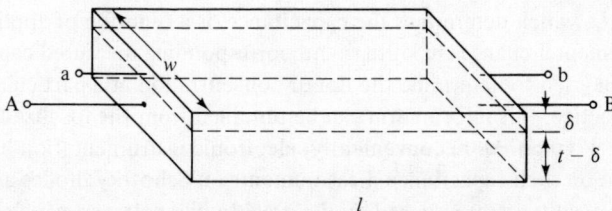

Fig. 2.11   Measurement of the doping profile.

23

where $\bar{\rho}_1$ is the average *resistivity* of the removed layer. Thus the resistivity and hence impurity concentration of the removed layer can be estimated from the measured values, since:

$$\bar{\rho}_1 = \frac{\rho_{s1}\rho_{s2}}{\rho_{s2} - \rho_{s1}} \delta \ \Omega\text{-m} \tag{2.7}$$

Thus the resistivity and hence the doping can be found by successive layer removal and resistance measurement.

An alternative method, which has the advantage of being non-destructive, is to evaporate a gold Schottky barrier diode of prescribed area on to the surface of the layer. Such diodes are similar in principle and construction to those described in section 5.5.4. The depletion capacitance of the diode, C, is then measured as a

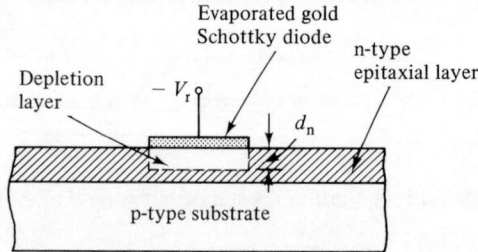

Fig. 2.12   Measurement of the impurity profile from the capacitance of a Schottky diode.

function of reverse bias voltage, V_r. The capacitance is given by the usual expression:

$$C = \left[\frac{e\epsilon_r\epsilon_0 N_d}{2} \right]^{1/2} V^{-1/2} \tag{2.8}$$

so that if C^{-2} is plotted as a function of V, the slope of the resulting linear graph provides an average value for the doping concentration in the depletion layer, N_d. This argument assumes a uniformly doped layer, whereas the value of N_d is often a function of distance into the layer. However, in such cases, since the depletion layer width, d_n, which determines the capacitance, is a function of applied voltage, incremental changes in voltage and corresponding measured capacitance changes can be used to determine the donor concentration at a particular depth, $N_d(d_n)$. In practice, this information can be obtained from the localized gradient of the C^{-2} vs. V graph. More conveniently, electronic instrumentation has been devised, based on such capacitance measurements on Schottky diodes and using small signal a.c. voltages superposed on d.c. reverse bias voltages, which provides direct information on the impurity profile.

24

Problems

2.1   The slope of the voltage-current curve obtained by a four-point probe measurement on a $12~\mu m$ thick n-type epitaxial layer is $800~\Omega$. Find (a) the sheet resistance (b) the resistivity, and (c) the approximate donor concentration of the layer. Assume an electron mobility of $0.13~m^2~V^{-1}~s^{-1}$. Use the answers to (a) to estimate the resistance along the length of an isolated strip of the layer which has surface dimensions $200 \times 10~\mu m$. How would the layer thickness normally be measured?

Ans.   (a) $3.62~k\Omega$/square, (b) $4.35 \times 10^{-2}~\Omega$-m, (c) $10^{21}m^{-3}$; $72~k\Omega$.

2.2   A four-point probe method is used to evaluate the doping profile of a p-type impurity in a silicon epitaxial layer. When the four equispaced probes are placed on the surface of the slice a probe current of $10~mA$ produces a voltage at the inner probes of $0.22~V$. What is the sheet resistance of the slice?

Successive layers, $1~\mu m$ thick, are to be removed from the epitaxial layer by anodic oxidation, monitoring the resistance by the four-point probe method, keeping the probe current fixed at $10~mA$. After the first layer is removed, the voltage measured is $0.55~V$. What is the mean resistivity and the average impurity concentration of the removed layer, assuming a hole mobility of $0.05~m^2~V^{-1}~s^{-1}$?

Explain the subsequent measurement procedure used to estimate the profile throughout the entire epi-layer.

Ans. $100~\Omega$/square; $1.7 \times 10^{-4}~\Omega$-m; $8 \times 10^{23}~m^{-3}$.

3. Oxide layer growth and window preparation

The production of silicon substrate slices supporting an epitaxial layer of doped silicon has been described in chapter 2. It is in this basic starting material that the components of monolithic integrated circuits are formed. The various circuit components are produced, as discussed in chapter 1, by diffusion of dopant impurities into the epi-layer. The areas over which particular diffusions are effective are defined by an oxide layer (which inhibits diffusion) with 'windows' cut in it, through which diffusion can take place. This chapter is devoted to the techniques used for the production of suitable oxide layers and for delineating diffusion windows in them.

3.1 Growth and properties of oxide layers on silicon

A layer of silicon dioxide (silica) is formed on the surface of a silicon slice by thermal oxidation at high temperatures in a stream of oxygen:

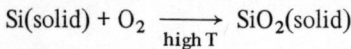

$$Si(\text{solid}) + O_2 \xrightarrow[\text{high T}]{} SiO_2(\text{solid})$$

The oxidation furnace used for this reaction is similar to the diffusion furnace shown in Fig. 4.15. The thickness of the oxide layer depends on the temperature of the furnace, the length of time that the slices are in it, and the flow rate of oxygen. The rate of oxidation can be increased by adding water vapour to the oxygen supply to the oxidizing furnace:

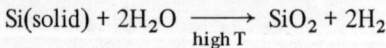

$$Si(\text{solid}) + 2H_2O \xrightarrow[\text{high T}]{} SiO_2 + 2H_2$$

Most 'thick' oxides are produced in this manner. The time and temperature required to produce a particular layer thickness are obtained from empirically determined design curves, of the type shown in Fig. 3.1. Layer thicknesses in the range 0.1–5 μm are commonly produced at temperatures between 1000 and 1200°C. A typical schedule might be to bubble oxygen through water into a furnace at 1100°C, to give an oxide growth rate of around 0.5 μm per hour.

A finished oxidized layer must be capable of completely masking the silicon slice from any dopant that is introduced in the subsequent diffusion stage; impurities can then only diffuse into the silicon slice through the carefully defined

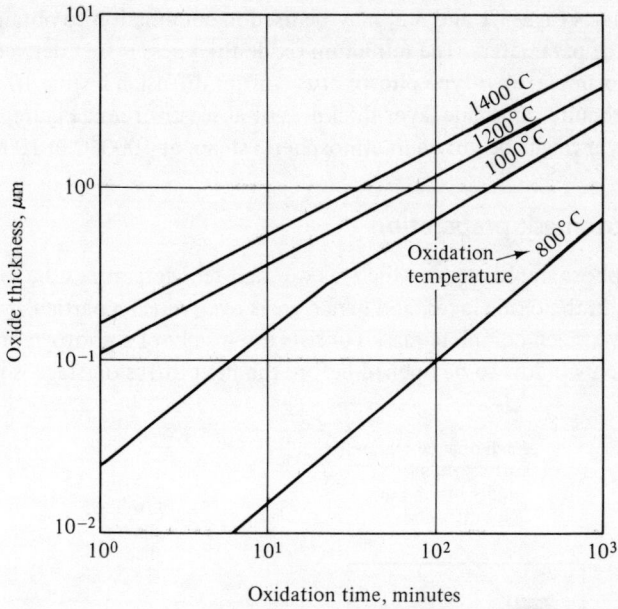

Fig. 3.1 *The rate of growth of silicon dioxide layers on silicon in atmospheric steam.*

windows that are cut in the oxide layer. The minimum thickness of oxide required for the complete masking of a particular dopant depends on the dopant and the diffusion parameters, e.g., time and temperature. The minimum thicknesses of silicon dioxide that are required for various diffusion temperatures and times can be obtained from empirical design curves, of the type shown in Fig. 3.2. The two

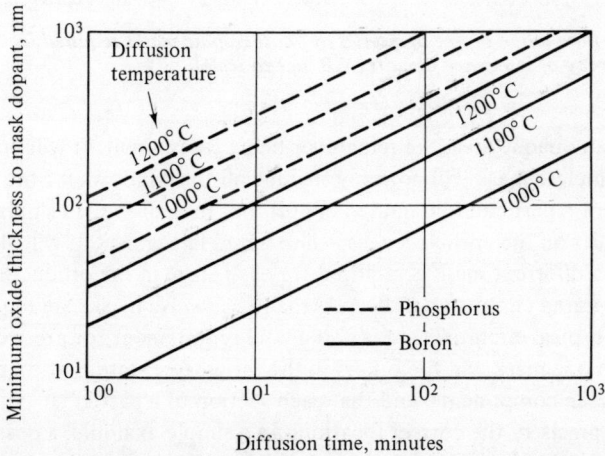

Fig. 3.2 *Minimum thickness of silicon dioxide required to mask a silicon slice from various dopants, under different diffusion conditions.*

27

design charts Figs. 3.1 and 3.2, may be used in conjunction to obtain a preliminary estimate of parameters. The minimum oxide thickness is first derived from Fig. 3.2; for example, an n-type phosphorus emitter diffusion lasting 10 minutes at $1000°C$ requires an oxide layer thickness of at least 0.1 μm. Figure 3.1 shows that such a layer might be grown in atmospheric steam at $1000°C$ in 10 minutes.

3.2 Photomask preparation

It is the photographic mask which, as we shall see, determines the location of all windows in the oxide layer, and hence areas over which a particular diffusion step is effective. Each complete mask consists essentially of a photographic plate on which each window to be opened before the next diffusion stage is represented,

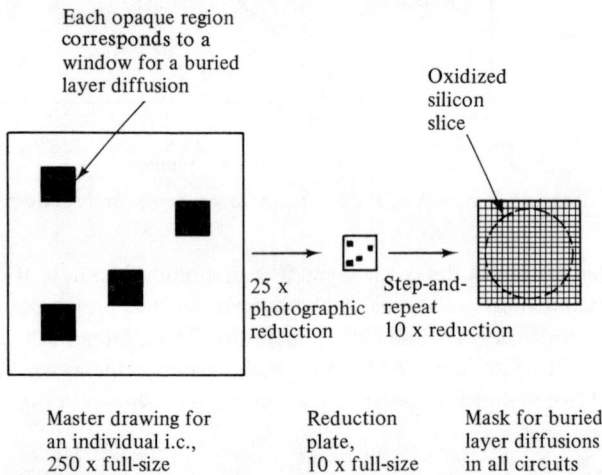

Each opaque region corresponds to a window for a buried layer diffusion

Oxidized silicon slice

25 x photographic reduction

Step-and-repeat 10 x reduction

Master drawing for an individual i.c., 250 x full-size

Reduction plate, 10 x full-size

Mask for buried layer diffusions in all circuits

Fig. 3.3 The production of one of a series of photographic masks required for the manufacture of an array of integrated circuits (N.B. not to scale).

full-size, by an opaque area, the remainder being transparent. It will be apparent that each complete mask will not only include all the windows for the production of one stage of a particular integrated circuit, but in addition, all similar areas for all such circuits on the entire slice, as is illustrated in Fig. 3.3. It will also be evident that a different mask is required for each stage in the production of an array of integrated circuits on a slice. As many as twelve masks are required for producing a typical circuit. There is also a vital requirement for precise registration between one mask and the others in the series, to ensure that there is no overlap between components, and that each section of a particular transistor, say, is formed in precisely the correct location. As a simple example, a possible set of masks which might be used to produce the elementary integrated circuit shown in Fig. 1.2 is illustrated in Fig. 3.4.

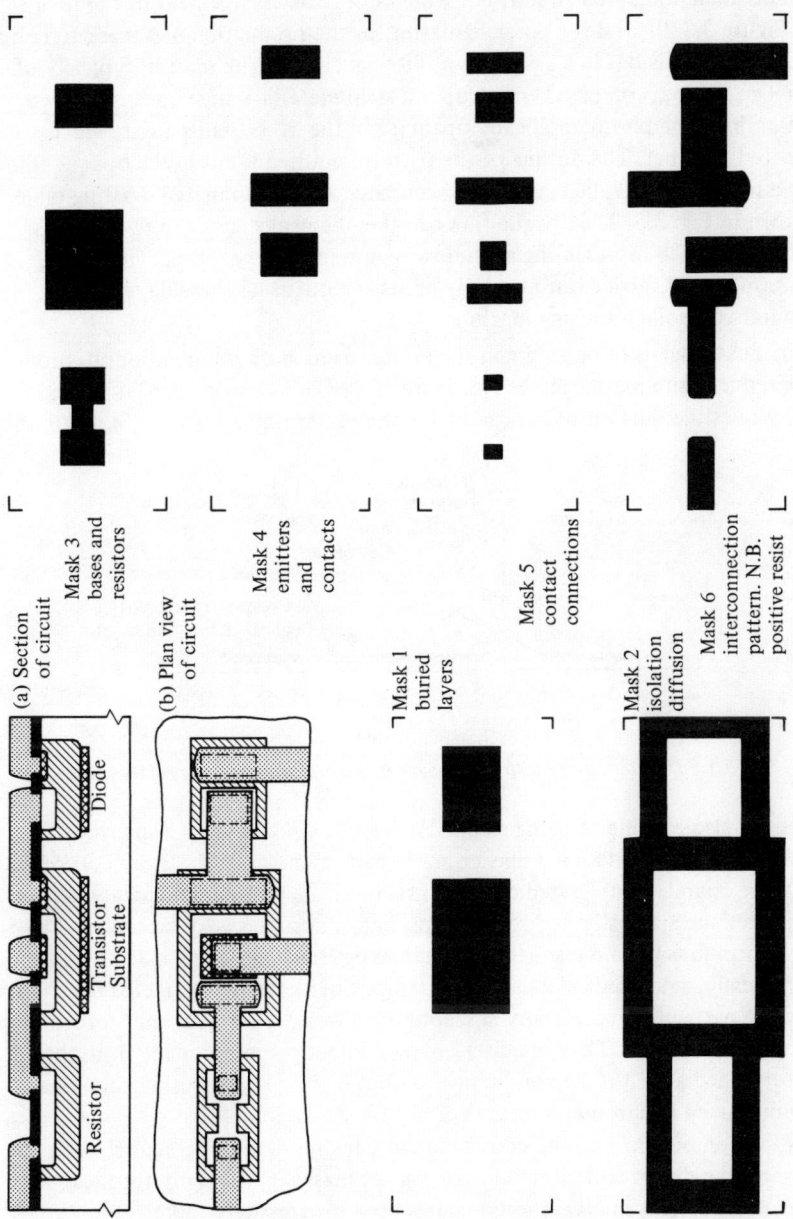

(a) Section of circuit

Resistor   Transistor   Diode
            Substrate

(b) Plan view of circuit

Mask 1 buried layers

Mask 2 isolation diffusion

Mask 3 bases and resistors

Mask 4 emitters and contacts

Mask 5 contact connections

Mask 6 interconnection pattern. N.B. positive resist

Fig. 3.4   A possible set of masks to produce a hypothetical integrated circuit (see also Fig. 1.2)

29

To make a mask for one of the production stages, a master is first prepared which is an exact replica of that portion of the final mask associated with one individual integrated circuit, but which is a 250x enlargement of the final size. Figure 3.3 illustrates a possible master for the production of a mask to define buried layer diffusions in a simple hypothetical circuit. The master, typically of order 1 m x 1 m, is prepared from cut-and-strip material which consists of two plastic films, one photographically opaque and the other transparent, which are laminated together. The outline of the pattern required is cut in the opaque film, using a machine-controlled cutter for accuracy, on an illuminated drafting table, as shown in Fig. 3.5. The opaque film can then be peeled away to reveal transparent areas, each representing a window region in the final mask. The cutting operation can be carried out manually or, as is more usual these days, by a computer-controlled drafting machine.

The next stage is to photograph the master using back illumination, to produce a 25x reduced sub-master plate. This plate is used in a step-and-repeat camera which serves the dual purpose of reducing the pattern by a further 10x to finished

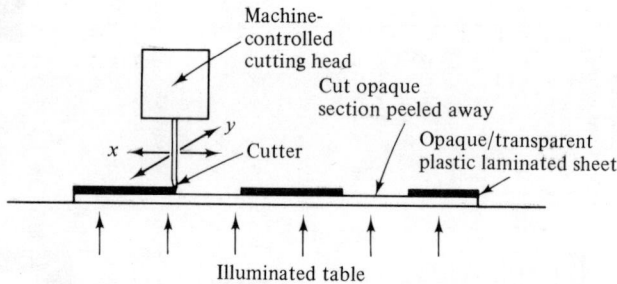

Fig. 3.5   *Production of masters for mask making, using cut-and-peel material.*

size and is also capable of being stepped mechanically to produce an array of identical patterns on the final master mask, each member of the array corresponding to one complete integrated circuit. Instead of the photographic plate being transported mechanically in discrete steps, better accuracy may be achieved by using continuous plate movement, discrete exposures then being made by an electronically synchronized xenon flash lamp which effectively freezes the motion. Because a particular mask is very susceptible to wear, it is useful only for a limited number of operations. Consequently, many contact prints are made from the mask obtained after the step-and-repeat photography. These working copies are often produced in chromium on glass slides for extra durability.

As the component packing density in complex modern integrated circuits has increased, the design and fabrication of master masks, particularly for interconnection patterns in MOST circuits, has been progressively more computer controlled. Designs for all basic electronic components are stored in a computer and the computer is programmed to determine mask sets appropriate to each stored component. The designer then inputs a preliminary layout design and the computer manipulates its stored masks to provide data for a set of mask patterns. The computer outputs its mask design on an x–y recorder for checking and provides a control tape for cutting of the mask masters. Alternatively, the computer output can be used to control the position and size of an electrically variable shutter in a pattern generator, to produce sub-master plates directly.

Electron beam machining of masks is also being developed. Very fine patterns can be produced by a computer controlled beam, which, because of its relatively low wavelength, can have extremely high resolution. Thus the final master mask can be produced directly, intermediate photographic reduction stages becoming redundant. Masks are made on chromium on glass slides, 50 mm square typically, using an electron beam of around 1 μm diameter, by means of an 'electron lithographic' technique. A chromium on glass slide is coated with an electron-sensitive resist film which is scanned with the required pattern by a computer-controlled electron beam. The electron-exposed resist is removed by a developer and the chromium film is etched away through the resist holes to complete the finished mask.

3.3 Photolithographic window preparation

Windows, through which subsequent diffusions are made, are opened in the oxidized silicon slice by the following technique. First, a photoresist light sensitive emulsion is placed on the surface of the oxidized silicon slice, either using an aerosol spray or an eye-dropper, while the slice is held on a vacuum chuck and rotated at high speed, typically 20 000 rev min^{-1}. This operation, done in a safe yellow light, produces a thin, uniform film of resist on the surface of the slice, as shown diagrammatically in Fig. 3.6(a). The slice is then baked, typically at 150°C, to remove solvents, again in a safe light. A photographic mask, produced as explained in the previous section, is accurately registered with the pattern on the slice produced by any previous operations, using a mask alignment jig, which is shown in rudimentary form in Fig. 3.6(b). The mask is then momentarily placed in contact with the photoresist film, still in the alignment jig, and exposed to ultraviolet light, Fig. 3.6(b). The photosensitive emulsion has the property of becoming insoluble in certain developers wherever it is polymerized by exposure to ultraviolet radiation. Unexposed areas under opaque parts of the mask can therefore be selectively removed with a photoresist developer, which is again sprayed on, Fig. 3.4(c). The remaining resist is hardened and acts as a convenient mask through which the oxide layer can be etched away, using buffered hydrofluoric acid, to expose areas of semiconductor underneath, Fig. 3.6(d). The remaining resist is finally removed with a mixture of sulphuric acid and hydrogen peroxide, and a final wash and dry stage completes the required window in the oxide layer.

The window opening operations described so far have employed *negative* photoresist emulsion. *Positive* emulsions are also available which are used in precisely the same manner, except that windows are opened wherever the mask

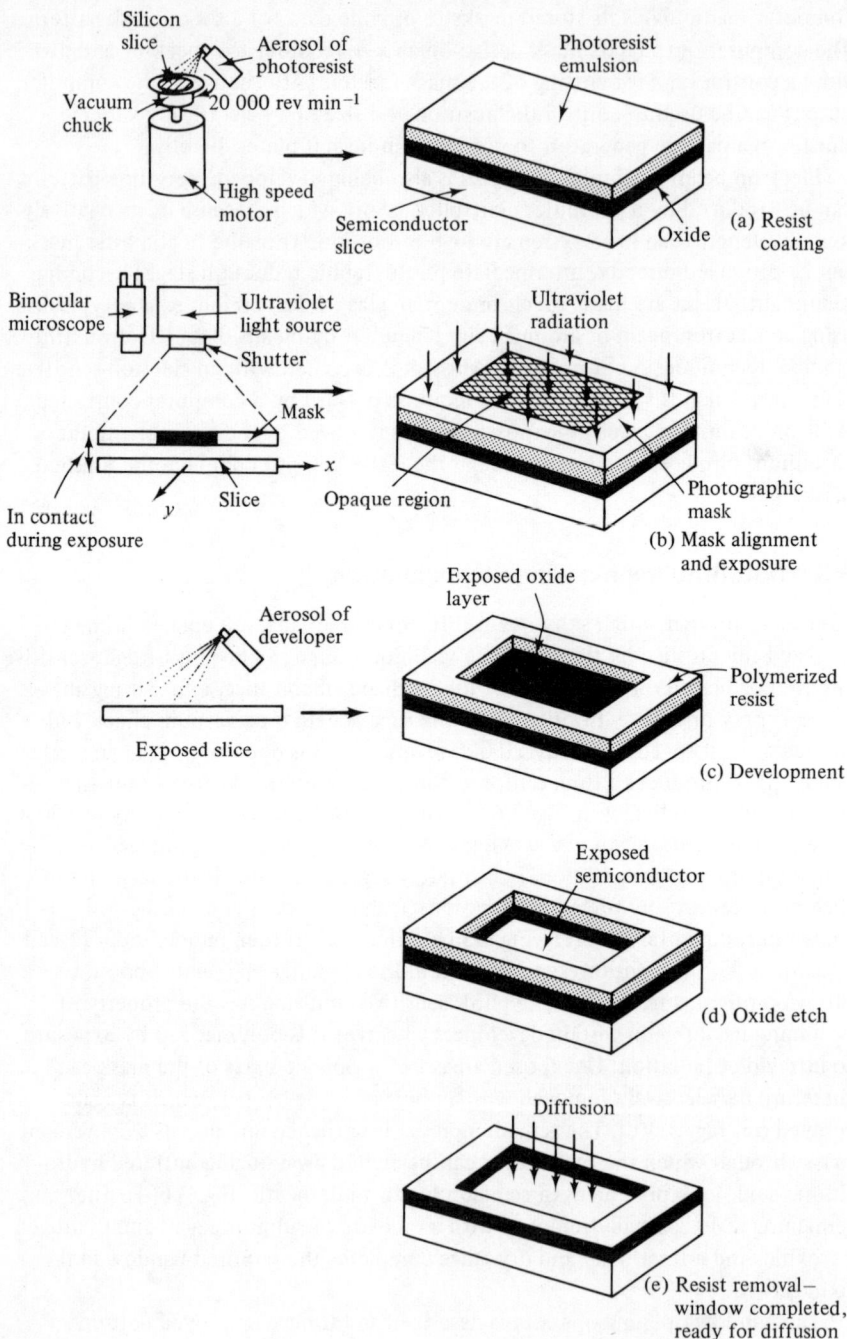

Fig. 3.6 *Preparation of windows in the oxide layer for the location of diffusions into a silicon slice.*

is transparent (rather than opaque as for the negative resist). This material is most useful when large areas of oxide are to be removed, for example for the definition of metallic interconnections, see Fig. 3.4.

Cleanliness at this and all other stages in the manufacturing process is essential. Operations are often carried out in laminar flow cabinets which are themselves housed in air-conditioned clean-rooms. Special clothing for all clean-room operatives is also obligatory.

4. Diffusion of dopant impurities

We have now described the stages in the production of monolithic epitaxial integrated circuits leading to the next process, in which circuit components are created in the silicon layer by successive diffusions of dopant impurities via windows in an oxide layer. In this chapter, we discuss the nature of the diffusion process and show how the properties of the doped layers can be controlled and predicted. Practical diffusion systems and the choice of suitable dopants for silicon integrated circuits will then be described.

4.1 Introduction

Most components in microcircuits are produced by a series of diffusion processes, in which controlled amounts of impurity dopant material are inserted into a silicon lattice. Impurity atoms are introduced on to the surface of a silicon slice and diffuse into the lattice because of their tendency to move from regions of high to low concentration. The diffusion process is therefore such that more dopant atoms are located near the surface of the slice than are inside the bulk of the material. This produces a resistivity profile which is a function of distance from the surface. Since the shape of the impurity profile influences the design and performance of a particular integrated circuit component, a reliable diffusion theory is required to predict the profile for a given set of operating conditions.

The material in which the diffusion takes place may be intrinsic, which is diffused locally with, say, donor atoms, to produce an n-type material. Much more commonly in integrated circuit processing, the material is already doped, say n-type, by a previous epitaxial or diffusion process; diffusion of a p-type impurity then creates a compensated p-type region.

4.2 The nature of impurity diffusion

Impurities move in the silicon lattice down a concentration gradient, by a series of random jumps. There are several possible mechanisms for this motion, dependent, among other things, on the particular dopant atom, but only those most pertinent to the common integrated circuit technology will be discussed.

4.2.1 Substitutional diffusion

In this process, impurity atoms diffuse by moving from a lattice site to a neighbouring one by substituting for a silicon atom which has vacated a usually occupied site, as shown diagramatically in Fig. 4.1. This diffusion mechanism is applicable to many of the most common dopants, such as boron, phosphorus, and arsenic.

In order for such an impurity atom to move to a neighbouring vacant site, it has to overcome a potential energy barrier of height E_{b1} eV, which is due to the breaking of covalent bonds. The probability of its having enough thermal energy to do this is proportional to an exponential function of temperature, $\exp(-E_{b1}/kT)$.

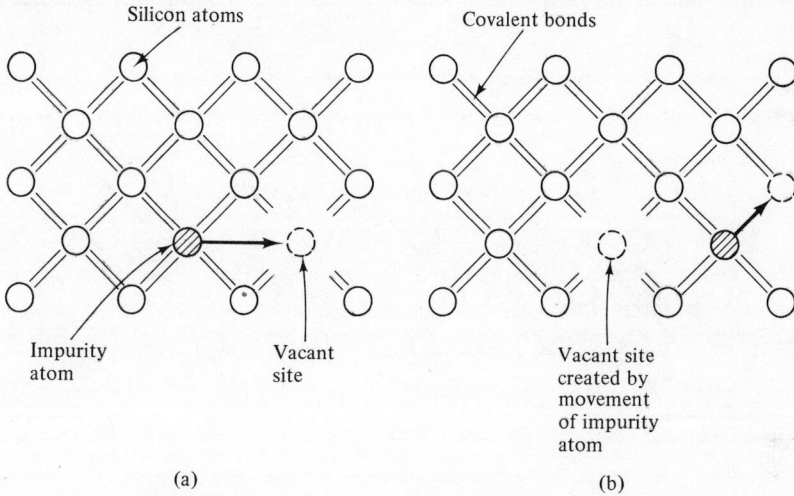

Silicon atoms   Covalent bonds

Impurity   Vacant   Vacant site
atom   site   created by
                              movement
                              of impurity
                              atom

(a)   (b)

Fig. 4.1   Two-dimensional representation of substitutional diffusion by dopants.

But whether it is able to move is also dependent on the availability of a vacant neighbouring site, and since an adjacent site is vacated by a silicon atom due to thermal fluctuation of the lattice, the probability of such an event is again an exponential function of temperature, $\exp(-E_v/kT)$, where E_v is the energy of formation of a vacancy in the lattice. Thus, if the frequency of lattice vibrations, or the rate at which they try to surmount the energy barrier is f_e, then the frequency with which impurity atoms move to neighbouring vacant sites f_n is

$$f_n = 4f_e \exp(-E_{b1}/kT) \exp(-E_v/kT) \qquad (4.1)$$

The factor 4 arises because of the presence of four neighbouring sites to accommodate possible moves. The energies in eq. (4.1) are typically of order 1 eV and the frequency f_e is around 10^{13} Hz, so the jump rate at ordinary temperatures is very slow, for example about 1 jump per 10^{50} years at room temperature! However,

35

the diffusion rate can be speeded up by an increase in temperature, and at temperatures around about those usually used, of order 1000°C, substantial diffusion of dopant impurities in sensible time scales can be practically realized.

4.2.2 Interstitial diffusion

This diffusion process is most applicable to the heavy metal ions, for example gold, which is introduced into silicon to reduce carrier lifetimes. Because of the relatively large size of such atoms, they do not usually substitute in the silicon lattice. However, there are five* interstitial voids present within the unit cell of the diamond lattice of the silicon, each of which is big enough to contain an impurity atom. An impurity atom located in one such void can move to a neighbouring void, as shown diagrammatically in Fig. 4.2. In doing so it again has to

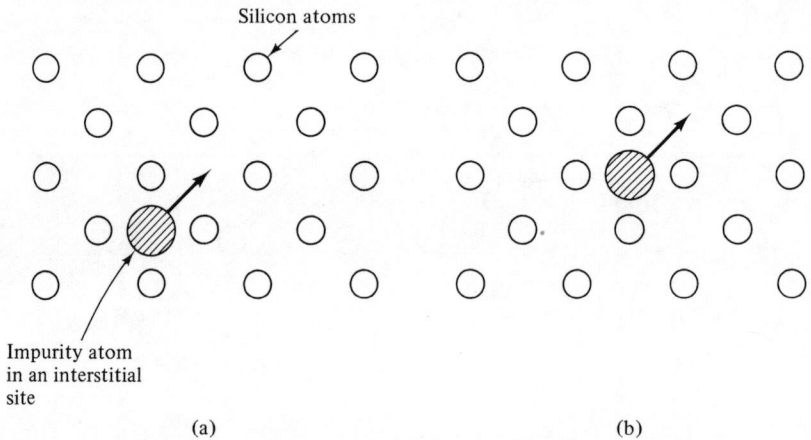

Silicon atoms

Impurity atom
in an interstitial
site

(a)   (b)

Fig. 4.2   Diffusion of dopant atom between interstitial sites.

surmount a potential barrier due to the lattice. This time, though, most neighbouring interstitial sites are vacant, so the frequency of movement is reduced to:

$$f_n = 4f_e \exp\left(-E_{B2}/kT\right) \tag{4.2}$$

Again the diffusion rate due to this process is very slow at room temperature but becomes practically acceptable at normal operating temperatures of around 1000°C. It will be noticed that the diffusion rate due to interstitial movements is much greater than for substitutional mechanisms. This is reflected in the respective diffusion coefficients for such processes, which will be discussed later.

---

* It is not readily apparent from the two-dimensional representation of the tetrahedral lattice that five voids exist; reference should be made to the three-dimensional diamond lattice structure to confirm that this particular number is present.

4.3 Diffusion of impurities in a concentration gradient

The speed at which impurities diffuse into a semiconducting lattice depends on the particular mechanism of diffusion and the temperature, as we have discussed. In addition the following factors are influential in determining the diffusion rate:

(a) the physical properties of a particular impurity, for example its atomic size, (b) the properties of the lattice environment, (c) the concentration gradient of impurities, and (d) the geometry of the parent semiconductor.

Diffusion can generally proceed in any direction, but for the planar geometry with which we are mostly concerned, consideration of motion in one dimension will usually be sufficient. In this case, the behaviour of diffusing particles is governed by *Fick's Law*, which relates the flux density of particles, j atoms m^{-2} s^{-1}, to the concentration gradient $\partial N/\partial x$ m^{-4}, N being the density of impurities, via a diffusion coefficient, D m^2 s^{-1}, as follows:

$$j = -D\frac{\partial N}{\partial x} \tag{4.3}$$

This equation implies that, for diffusion processes, the particle flux density, or rate of flow of particles per unit area, is directly proportional to the concentration gradient. The negative sign indicates that diffusion occurs away from regions of high concentration, i.e., in the opposite direction to increasing concentration gradient, as shown in Fig. 4.3. In general the dopant impurities are not charged, nor do they move in an electric field, so the usual drift mobility term, which is associated with eq. (4.3) when it applies to electrons or holes, can be omitted.

If we consider the small element of semiconducting solid shown in Fig. 4.4(a) then, provided no atoms are created or annihilated in the volume, the change in the number of atoms contained, per unit area and time, is equal to the difference between the particle fluxes in and out, or

$$\frac{\partial \bar{N}}{\partial t} \cdot \frac{\mathrm{d}xA}{A} = j(x) - j(x + \mathrm{d}x)$$

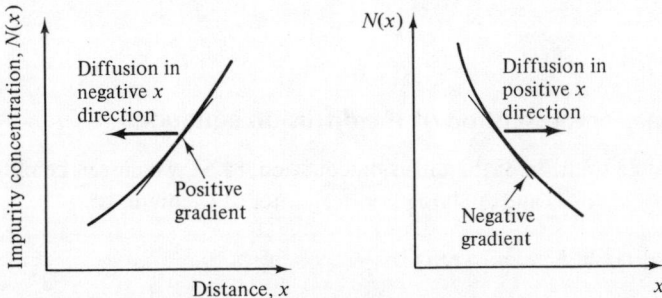

Fig. 4.3   Diffusion of impurities.

37

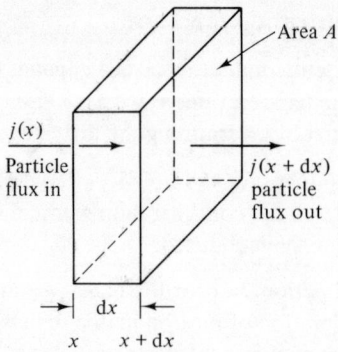
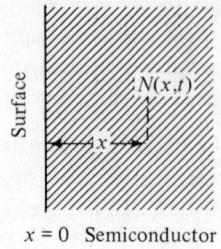

Fig. 4.4   (a) The elemental volume used to derive the transport equation and (b) the transport equation applied to planar diffusion.

where $\bar{N}$ is the average particle density. As dx approaches zero and $\bar{N}$ goes to $N(x)$ in the limit, then the equation reduces to:

$$\frac{\partial N}{\partial t} = -\frac{\partial j}{\partial x} \tag{4.4}$$

which is called the one-dimensional *transport equation.*

The expression for the particle flux density, j, as given by eq. (4.3), can now be substituted to give the *diffusion equation*, sometimes also known as Fick's second law:

$$\boxed{\frac{\partial N}{\partial t} = D\frac{\partial^2 N}{\partial x^2}} \tag{4.5}$$

The solution of this equation gives the impurity concentration, N, at some distance x from the origin, usually the surface of the semiconductor, as shown in Fig. 4.4(b), at time t.

4.4 The general solution of the diffusion equation

One possible solution of the diffusion equation, (4.5), which can be obtained by separating variables and applying Fourier's integral theorem, is:

$$N(x, t) = t^{-1/2} \exp\left[-x^2/(4Dt)\right] \tag{4.6}$$

That this is a solution is most easily confirmed by substitution.

Since eq. (4.6) represents only one particular integral of the diffusion equation, a family of such solutions can be devised, and since the equation is linear, the sum of any number of such integrals is also a solution, viz.,

$$N = \sum_n \frac{1}{2\sqrt{(\pi Dt)}} a_n \cdot \exp\left[-(x-x')^2/(4Dt)\right] \tag{4.7}$$

where x' is a series of arbitrary constants. Since the choice of x' is arbitrary, the summation over discrete values of x' can be replaced by an integral to give:

$$N(x, t) = \frac{1}{2\sqrt{(\pi Dt)}} \int_{-\infty}^{+\infty} f(x') \, e^{-(x-x')^2/(4Dt)} \, dx' \tag{4.8}$$

This is the general solution of the diffusion equation for diffusion in an infinite body. It can be shown that $f(x')$ corresponds to the initial concentration distribution of diffusants at time $t = 0$.

We next consider more specific solutions of the diffusion equation which are directly applicable to the planar diffusion processes used in integrated circuit technology.

4.4.1 Planar diffusion from a constant source of dopants

This diffusion technique is one of the most commonly occurring processes in planar technology. Dopants are introduced on to the surface of a hot silicon slice and are allowed to diffuse into the material, the amount of dopant at the surface being maintained at a constant level throughout. Such a *constant source* diffusion is illustrated in Fig. 4.5(a).

The solution to the diffusion equation which is applicable in this situation is most easily obtained by first considering diffusion totally inside a material in which the initial concentration changes abruptly in some plane at $x = 0$, from N_0

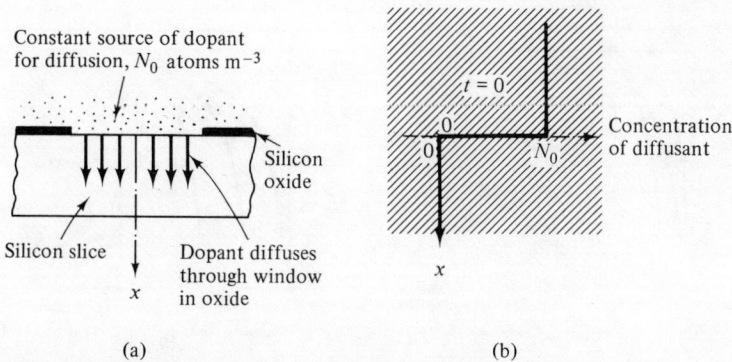

Fig. 4.5 Constant-source diffusion. (a) a conceptual view of diffusion in planar geometry, (b) initial conditions assumed to deduce the impurity profile, $N(x, t)$.

to zero, as shown in Fig. 4.5(b). With these boundary conditions, the general solution to the diffusion equation, eq. (4.8), becomes:

$$N(x, t) = \frac{N_0}{2\sqrt{(\pi Dt)}} \int_{-\infty}^{0} e^{-(x-x')^2/(4Dt)} \, dx' \tag{4.9}$$

The solution of this equation is

$$N(x, t) = \frac{N_0}{2}\left[1 - \mathrm{erf}\left\{\frac{x}{2\sqrt{(Dt)}}\right\}\right] \tag{4.10}$$

where

$$\mathrm{erf}\left\{\frac{x}{2\sqrt{(Dt)}}\right\}$$

represents the *error function*, which is tabulated. Alternatively, the solution can be expressed in terms of a tabulated *complementary error function*, as follows:

$$N(x, t) = \frac{N_0}{2}\,\mathrm{erfc}\left\{\frac{x}{2\sqrt{(Dt)}}\right\} \tag{4.11}$$

This represents the impurity profile in an assumed hypothetical infinite body; the concentration of diffusants at various times is as shown in Fig. 4.6(a).

In the practically occurring planar situation shown in Fig. 4.5(a), the surface concentration remains constant at all times during the diffusion, at N_0. The dis-

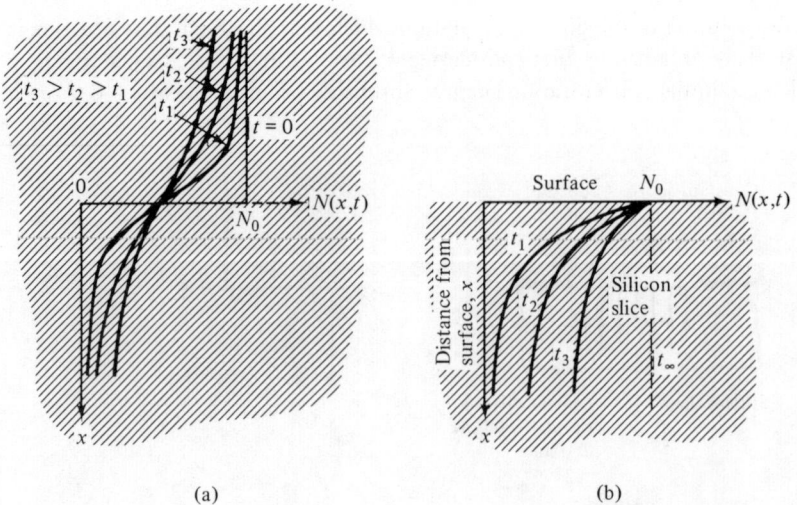

(a)   (b)

Fig. 4.6   Impurity concentrations in (a) a hypothetical infinite solid and (b) a planar constant-source diffusion.

40

tribution of impurities in this case, $N(x, t)$, can be deduced from eq. (4.11) and is given by:

$$N(x, t) = N_0 \, \text{erfc} \left\{ \frac{x}{2\sqrt{(Dt)}} \right\} \qquad (4.12)$$

The change in concentration of diffusing dopants with time, as described by this equation, is then as shown in Fig. 4.6(b). The surface concentration is always held at N_0, falling to some lower value away from the surface. If a sufficiently long time is allowed to elapse, it is possible for the entire slice to acquire a dopant level of $N_0 \, \text{m}^{-3}$. Normalized design charts can be constructed, as shown in Fig. 4.7, which allow eq. (4.12) to be solved graphically, so that the impurity profile can be deduced for any particular diffusion time and under particular conditions of

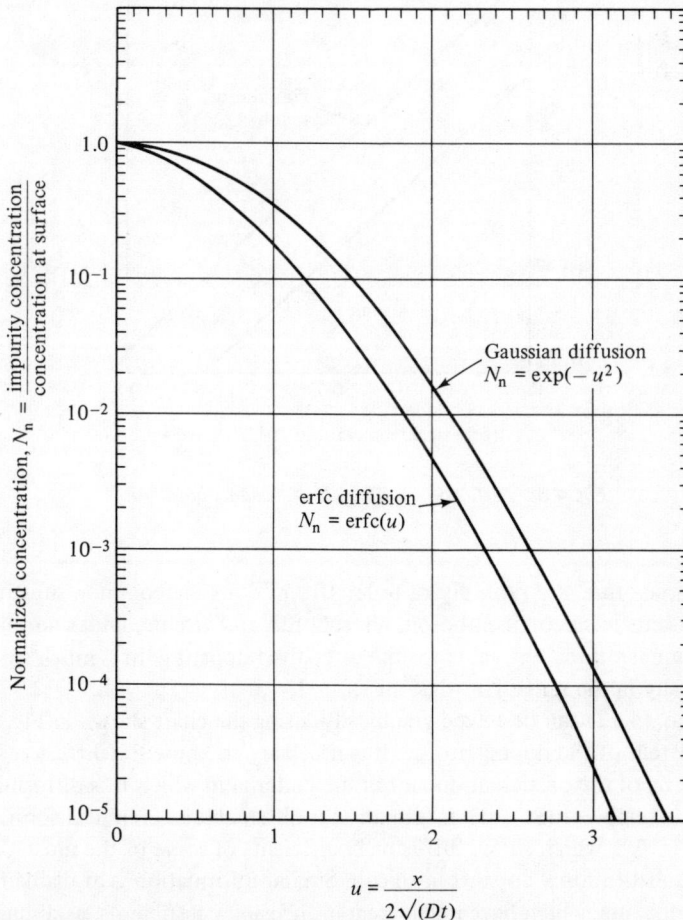

Fig. 4.7 *Design charts used to predict impurity profiles for erfc and Gaussian diffusions.*

41

temperature, which affects the diffusion coefficient D, and surface concentration N_0.

Because of the high surface concentrations which are possible using *constant source* or *complementary error function* diffusion, the technique is particularly useful for emitter type diffusions, as we shall see later.

There is obviously an upper limit to the concentration of any impurity that can be accommodated in the lattice of a particular solid at some temperature, T. This maximum concentration, which usually determines the surface concentration in a constant-source diffusion, is called the *solid solubility* of the particular impurity. Values of the solid solubility, which are commonly tabulated as func-

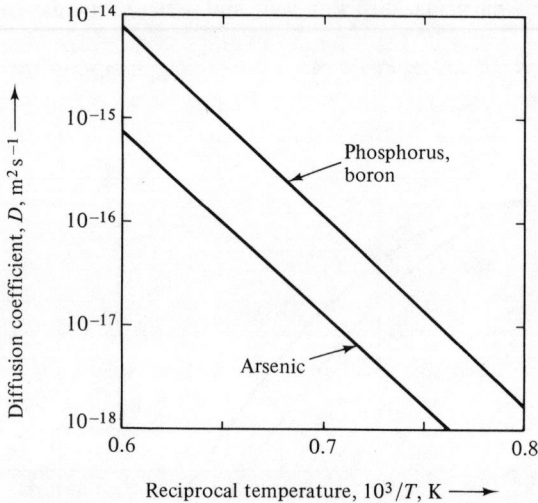

Fig. 4.8 Diffusion coefficients of dopants in silicon.

tions of temperature, are typically of order $10^{27}\,m^{-3}$ for the common substitutional diffusants in silicon, i.e., boron, phosphorus, and arsenic, under normal operating temperatures; the value for the interstitial dopant gold is much smaller, being typically in the range 10^{22}–$10^{23}\,m^{-3}$.

Before eq. (4.12) can be solved graphically, using the chart shown in Fig. 4.7 to obtain details of the doping profile, it is necessary to know the diffusion coefficient, D, of the particular dopant in the material in which it is diffusing, at the diffusion temperature. Such information is obtainable in graphical form, as shown in Fig. 4.8, which shows diffusion coefficients of some of the most important substitutional dopants in silicon. Similar information is available for the interstitial dopants, which have much greater diffusion coefficients, as discussed; for example, for gold diffusing in silicon at 850°C, $D \approx 10^{-11}\,m^2\,s^{-1}$.

42

4.4.2 Planar diffusion from a limited source of dopants

We have seen that an error function diffusion is ideal for relatively shallow high-conductivity regions, such as transistor emitters. However, for other portions of an integrated circuit, for example for transistor bases, a more uniform distribution than can be provided by an erfc diffusion is required. Such impurity profile levelling can be achieved by terminating the supply of diffusant after some time, while keeping the semiconductor at a high temperature. An initial diffusion, often erfc type, results in a finite quantity of diffusant, Q atoms m^{-2}, being deposited in a very thin layer on the surface as shown in Fig. 4.9(a). The supply of dopant is then turned off and diffusion thereafter continues from the fixed expendable source that has already been diffused into the surface layer. Such a technique is often aptly called a *limited source* diffusion.

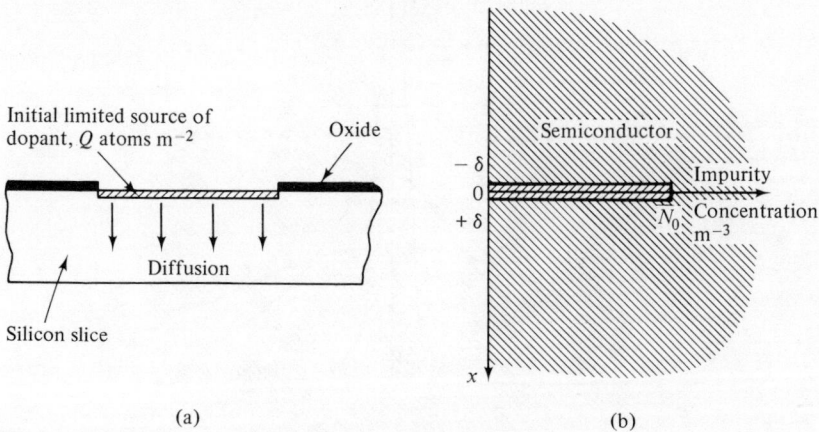

Fig. 4.9 (a) Limited source diffusion, (b) mathematical model used to determine the impurity profile.

Derivation of the profile for such a diffusion is again most easily achieved by considering initially the hypothetical but parallel case of a thin layer of impurities embedded in an infinite semiconductor as shown in Fig. 4.9(b). The impurities, totalling Q per m^2, or alternatively of density N_0 m^{-3}, are assumed to be initially located in a layer, thickness 2δ, centred at $x = 0$, as shown. Substituting such initial boundary conditions into the general solution of the diffusion equation, eq. (4.8), gives the density of dopant distant x from the origin at time t as

$$N(x, t) = \frac{N_0}{2\sqrt{(\pi D t)}} \int_{-\delta}^{\delta} e^{-(x-x')^2/(4Dt)} \, dx' \tag{4.13}$$

Remembering that $N_0 = Q/2\delta$ and changing the limits gives:

$$N(x, t) = \frac{Q}{2\delta \sqrt{(\pi D t)}} \int_{0}^{\delta} e^{-(x-x')^2/(4Dt)} \, dx' \tag{4.14}$$

43

Carrying out the integration then leads to:

$$N(x, t) = \frac{Q}{2\sqrt{(\pi D t)}} \exp\left(-x^2/4Dt\right) \qquad (4.15)$$

The distribution of impurities after various times, as predicted by this equation, would then be of the form shown in Fig. 4.10.

In the practical planar situation, instead of diffusion being possible in both directions, as considered in the model, the limited amount of diffusant is initially at the surface of a slice and can only diffuse in one direction, i.e., into the slice. The distribution of dopant is described by eq. (4.15) for two-way diffusion, but

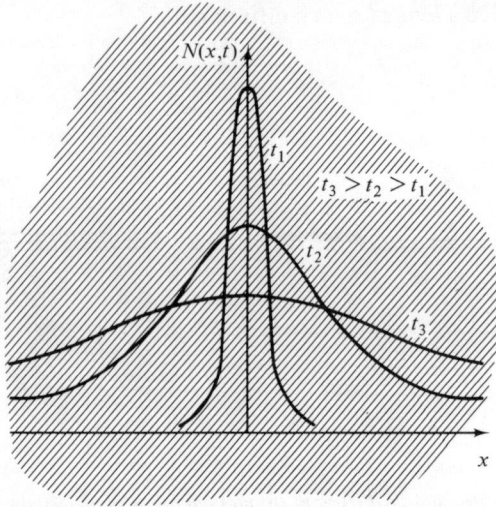

Fig. 4.10 *Impurity profiles for a limited source of dopant located at x = 0 inside the body of a semiconductor.*

if diffusion is constrained to the +x direction only, as in practice, the impurity concentration at any point and at a particular time is doubled, which results in

$$N(x, t) = \frac{Q}{\sqrt{(\pi D t)}} \exp\left(-x^2/4Dt\right) \qquad (4.16)$$

This is a Gaussian distribution and the process is sometimes called, alternatively, a *Gaussian diffusion*. Impurity profiles as a function of time can be deduced from eq. (4.16) and are of the general form shown in Fig. 4.11. Alternatively, this information can be obtained quantitatively using design curves based on eq. (4.16), such as in Fig. 4.7. An essential difference between the two types of diffusion technique so far discussed is illustrated by Fig. 4.11; whereas the surface concentration is held constant for an error function diffusion, it decays with

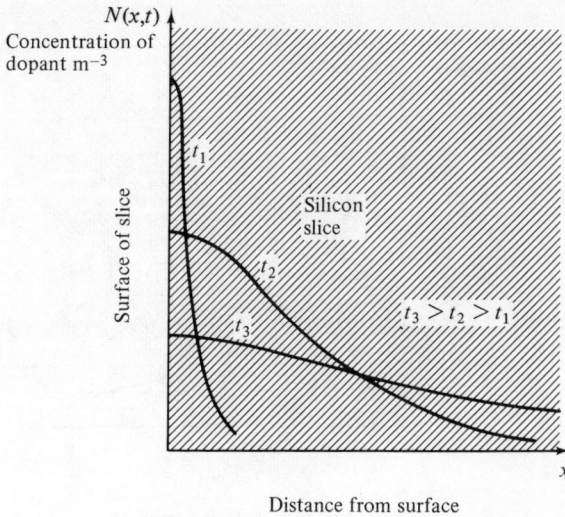

Fig. 4.11   Impurity profiles for a limited source diffusion in a planar geometry.

time for the Gaussian type. Hence a Gaussian diffusion is best suited to deep lying, fairly uniform layers in which a low surface concentration is acceptable, for example, for transistor bases.

4.4.3 Two-step diffusion

Two-step diffusion, in which the impurity concentration and profile can be carefully controlled and the type of diffusion, whether erfc or Gaussian, determined by the choice of operating conditions, is one of the most widely used techniques presently employed.

In step one, the *deposition stage*, a constant source diffusion is carried out for a short time, usually at a relatively low temperature, say $1000°C$. In step two, the *drive-in stage*, the impurity supply is shut off and the existing dopant is allowed to diffuse into the body of the semiconductor, which is now held at a different temperature, say $1200°C$, in an oxidizing atmosphere. The oxide layer which forms on the surface of the slice during this stage prevents further impurities from entering, or those already deposited, from diffusing out.

The final impurity profile is a function of diffusion conditions, such as temperature, time, and diffusion coefficient, for each step, as shown in Fig. 4.12. An error function type of distribution results if

$$D_1(T_1) . t_1 \gg D_2(T_2) . t_2$$

but the distribution will be Gaussian if the converse inequality is true. Here the subscripts refer to steps 1 and 2, and it should be noted that $D_1 \neq D_2$, even

45

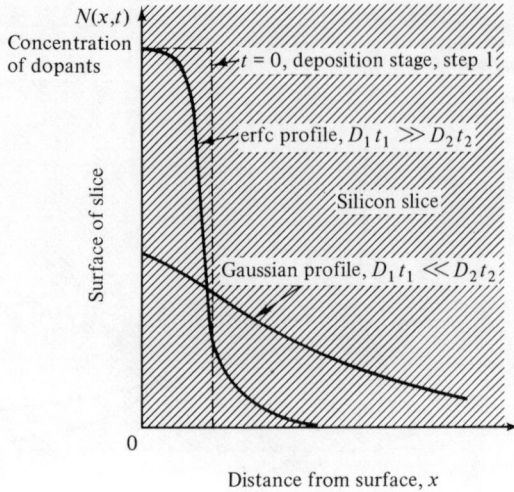

$N(x,t)$
Concentration
of dopants

$t = 0$, deposition stage, step 1

erfc profile, $D_1 t_1 \gg D_2 t_2$

Silicon slice

Surface of slice

Gaussian profile, $D_1 t_1 \ll D_2 t_2$

0

Distance from surface, x

Fig. 4.12   Two-step diffusion profiles.

though the dopant is unchanged, because of the different temperatures of each step.

Often the deposition stage results in surface concentrations which approach the solid solubility limit corresponding to the dopant and the temperature.

4.5 Junction formation

In planar integrated circuit technology, dopants are often introduced by diffusion into a semiconductor which already has a background concentration of impurities of opposite sign. For example, a p-type diffusion may be made into an n-type epitaxial layer which has a uniform donor concentration of N_d, say, as shown in Fig. 4.13. A p–n junction then occurs at the point where the p-type impurity concentration just equals that of the n-type background, i.e., at the position where full compensation occurs, and the net effective impurity concentrations are as shown. The distance that the junction is located from the surface, x_j, can thus be estimated by equating N_d to the $N(x, t)$ expression appropriate to the type of diffusion, evaluated at $x = x_j$. For example if the particular diffusion considered were a limited source type, with Gaussian profile described by eq. (4.16), then, at the junction

$$N_d = N(x_j, t) = \frac{Q}{\sqrt{(\pi D t)}} \exp\left(-x_j^2/4Dt\right) \tag{4.17}$$

from which x_j could be estimated.

In monolithic integrated circuit structures, the components are synthesized as we have discussed, by successive diffusions of donor and acceptor impurities, to

46

produce alternate layers of n- and p-type semiconductor. Consider, for example, the planar transistor shown diagrammatically in Fig. 4.14. The collector is usually formed in the fairly uniformly doped n-type epitaxial layer. Diffusion of acceptor impurities into the layer then creates the p-type base region, as just described. Finally, a second diffusion of donors reconverts the p-type material near the surface to n^+-type, to create the transistor emitter. Since the surface concentration of the first p-type base diffusion must be less than for the n^+ emitter diffusion but its depth has to be greater, this diffusion is Gaussian in character, whereas

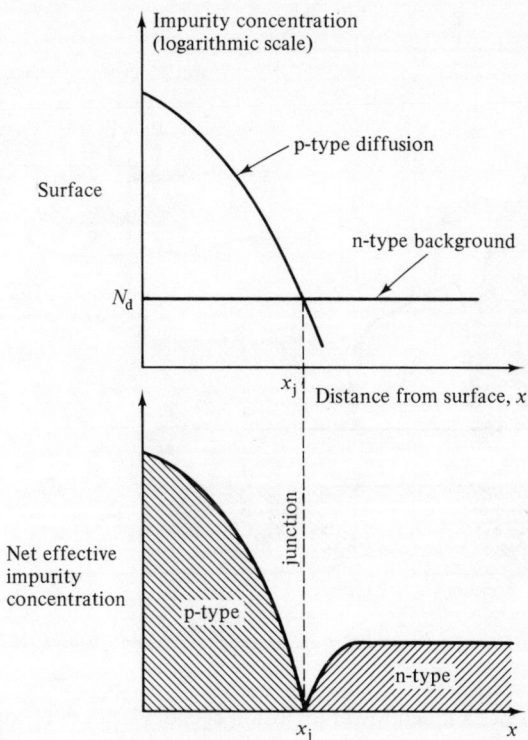

Fig. 4.13   Junction formation by a compensation diffusion.

the second, n^+-type, emitter diffusion is an erfc type. The values shown in Fig. 4.14 are typical for a high-speed switching transistor.

It is evident that the same region of a slice of semiconductor might be subject to several diffusion processes. Dopant types must be so chosen that the first diffusion takes place at a relatively high temperature with low diffusion coefficient material. If any subsequent shallower diffusion is then arranged to occur at a lower temperature with a higher diffusion coefficient dopant, the impurities deposited previously do not continue to diffuse appreciably during the second process. For example, arsenic is often chosen as a dopant for the n^+ buried layer

47

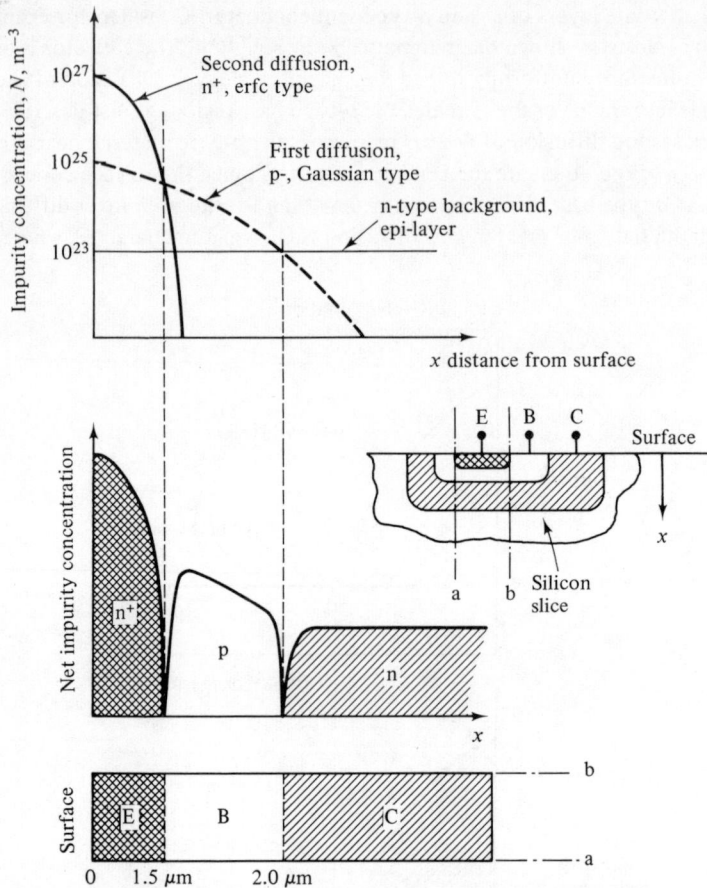

Fig. 4.14   Fabrication of a planar I.C. transistor by multiple diffusions (N.B. not to scale).

diffusion; since it has a much lower diffusion coefficient than boron or phosphorus dopants it remains in position during the subsequent operations in which these materials are diffused.

4.6 Practical diffusion systems

4.6.1 *p*-type impurities

Typical possible *p*-type diffusants in silicon are gallium, aluminium, boron, and indium. In practical diffusion systems the choice of dopant is much more restricted than might be supposed from this list. For example, gallium can be discounted since it has a relatively large diffusion coefficient in silicon dioxide, and the usual oxide window-opening technique for locating diffusions would be inoperative. Similarly, indium is of little interest in this context because of its high acceptor level of 0.16 eV, compared with 0.01 eV for boron, which indicates that not all such acceptors would be ionized at room temperature to produce a hole. The disadvantage with aluminium is that it reacts strongly with any oxygen that is present in the silicon lattice.

For these reasons, boron is an almost exclusive choice as an acceptor impurity in silicon. It has a moderate diffusion coefficient, typically of order 10^{-16} m^2 s^{-1} at 1150°C (see Fig. 4.8), which is convenient for precisely controlled diffusions.

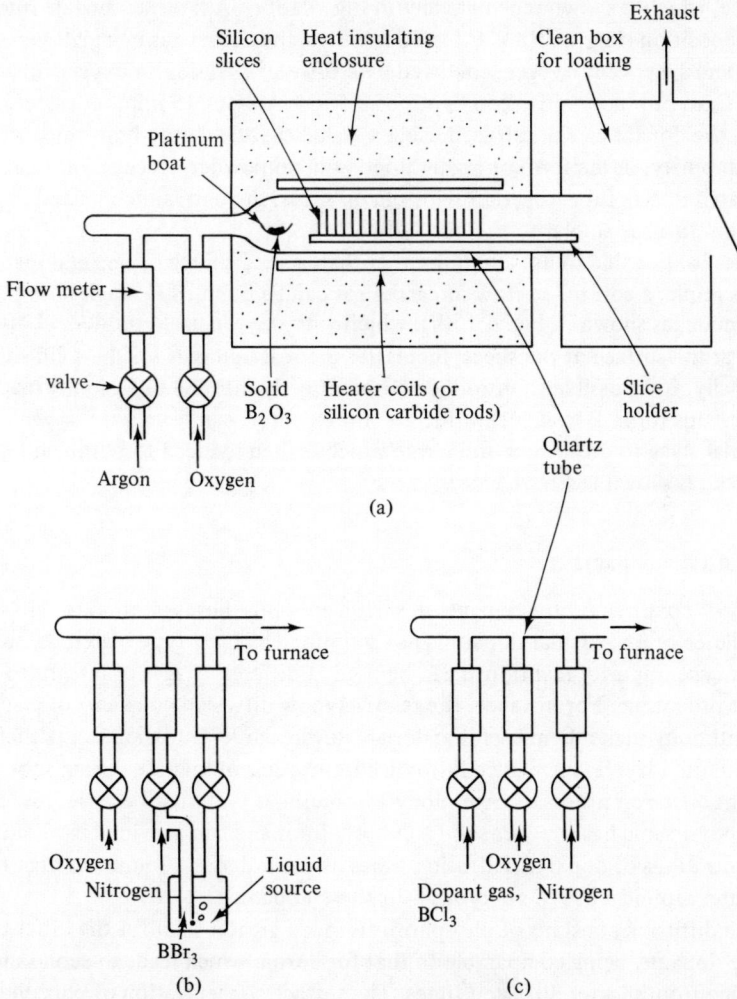

Fig. 4.15   *Electric diffusion furnace for boron dopant (a) solid source, (b) liquid source, and (c) gaseous source.*

It has a solid solubility limit of around 5×10^{26} atoms m^{-3}, so that surface concentrations can be widely varied, but most reproducible results are obtained when the concentration is approximately 10^{24} m^{-3}, which is typical for transistor base diffusions.

In most systems, a surface reaction between boron trioxide and silicon provides boron at the surface of the semiconductor, the reaction at diffusion temperatures being:

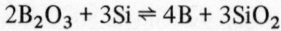

$$2B_2O_3 + 3Si \rightleftharpoons 4B + 3SiO_2$$

This interaction and the subsequent diffusion are carried out in a resistance furnace, which is shown schematically in Fig. 4.15(a). A typical schedule might be a deposition stage at $1000°C$ for 30 minutes, the carrier gas being 99 per cent argon and 1 per cent oxygen, followed by a drive-in diffusion in oxygen at $1100°C$, to produce a diffusion depth of 0.5 μm in about 15 minutes. Alternatively, the dopant can be diffused using a sealed capsule method, in which the solid impurity, in the form of highly doped silicon powder, is sealed into an evacuated quartz tube, together with silicon slices; the unit is then heated in a diffusion furnace similar to the one shown.

It is also possible to use a liquid rather than a solid source of boron dopant. For example, a controlled flow of carrier gas can be bubbled through boron tribromide, as shown in Fig. 4.15(b), which with oxygen again produces boron trioxide at the surface of the slices; thereafter the reaction is as for the solid source.

Finally, it is possible to introduce boron directly into the diffusion furnace in a gaseous form. Boron trichloride gas, for example, can react with oxygen in an initial stage to produce boron oxide which is then reduced to boron and is diffused, as shown in Fig. 4.15(c).

4.6.2 n-type dopants

The most common n-type dopants in silicon are phosphorus, antimony, and arsenic. The choice of a particular dopant is not so limited as for p-type materials, and each n-type impurity mentioned can be used at different stages of integrated circuit processing. For instance, the relatively low diffusion constants of arsenic and antimony make them useful materials for the earlier diffusion stages such as for n^+ buried layers, since once introduced, they do not migrate during subsequent diffusion processes. Antimony is sometimes preferred because it is less toxic but arsenic has a higher solid solubility limit and can provide bigger surface concentrations of dopants. Diffusion times of the order of 10 hours at high temperatures around $1200°C$ are typical for these dopants.

The diffusion constant of phosphorus is much greater than for the other two n-type dopants, being comparable to that for boron, which leads to economies resulting from shorter diffusion times. The surface concentration of phosphorus dopant is again variable over a wide range but is typically 10^{26} m^{-3} for n^+ erfc emitter diffusions.

Practical systems used to introduce n-type impurities on to slices for diffusion are similar to those described in the previous section. For example, a phosphorus source can be solid phosphorus pentoxide, liquid phosphorus oxychloride, or gaseous phosphine (PH_3). The latter, being toxic and explosive, needs some care in handling. It might be noted here that because of the precise control that is necessary during the various diffusion processes, and because of inherent difficulties of reproducibility due to contamination, it is customary for a separate diffusion furnace to be provided for each stage. A bank of such diffusion furnaces is installed for sequential slice processing, a separate furnace being used for each of the following steps: oxidation, buried layer (arsenic) diffusion and drive-in, isolation (boron) diffusion deposition, drive-in, base deposition (boron), drive-in, emitter deposition (phosphorus), drive-in.

4.6.3 Diffusion of interstitial dopants

Gold is often diffused into silicon circuits to enhance the recombination rate and so increase the switching speed of active devices, as will be discussed later.

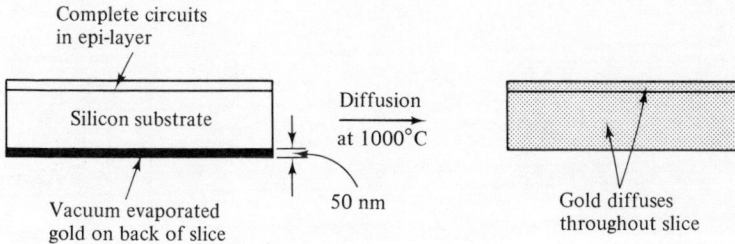

Fig. 4.16 Diffusion of gold.

Since gold has the relatively large diffusion coefficient associated with interstitial diffusants, a gold diffusion is often the last slice processing stage and takes place at a relatively reduced temperature. Because of the difficulty in controlling the gold impurity profile, it is usual to coat the back of the entire slice, using vacuum evaporation techniques, and to diffuse the impurity throughout the whole slice to a uniform level, as illustrated in Fig. 4.16.

Problems

4.1 A boron diffusion is to be made into an n-type, 0.005 Ω-m, silicon epitaxial layer. The impurity concentration at the silicon surface is maintained at 5×10^{25} atoms m^{-3} throughout. The diffusion takes place at 1100°C, at which temperature the diffusion coefficient of boron is 4.3×10^{-17} m^2 s^{-1}. A junction is to be formed 5 μm below the surface. For what length of time should the diffusion be carried out? An electron mobility in silicon of 0.15 m^2 V^{-1} s^{-1} may be assumed.

Ans. 5.7 h.

51

4.2   A particular boron drive-in diffusion into silicon lasts for 1 hour at a temperature of 1150°C; the diffusion coefficient of boron impurities in silicon at this temperature is 10^{-16} m^2 s^{-1}. The background concentration of donor impurities in the silicon before drive-in is 10^{22} m^{-3} and the junction depth afterwards is 1 μm. What concentration of boron atoms was pre-deposited on the surface of the silicon slice before drive-in?

Suggest how the pre-deposition stage might be accomplished.

Ans. 2.1×10^{16} m^{-2}.

4.3   An isolation diffusion is made through an n-type epitaxial layer, 10 μm thick, having an impurity concentration of 10^{22} atoms m^{-3}. The effective acceptor concentration at the surface of each diffused region is 5×10^{25} atoms m^{-3} and is maintained constant throughout the diffusion. Calculate the minimum time for the diffusion at 1200°C when the diffusion coefficient has a value 3.0×10^{-16} m^2 s^{-1}. How could this time be reduced?

Ans. 3.4 h.

5. Integrated circuit components

The overall manufacturing process used to produce monolithic integrated circuits has already been discussed briefly in chapter 1 and subsequent chapters have described the fabrication procedures in more detail. We now turn our attention to the various types of integrated component that can be made using the processes outlined. It is important that circuit designers as well as manufacturers are aware of the types of passive and active device available for integration, together with their characteristics and circuit limitations.

It will be noted, from earlier discussions on the general principles of monolithic integrated circuit fabrication, that all components are formed by a series of simultaneous, identical diffusion processes; for example all monolithic resistors are formed under the same diffusion conditions and at the same time as transistor bases. This not only limits the choice of possible components but also reduces flexibility in their design. For example, a typical base diffusion might have a sheet resistivity of 100 Ω/square, which constrains the design of resistors which utilize this layer. It is usual for the diffusion profiles and surface concentrations of the various constituent layers of planar integrated circuits to be chosen so as to optimize the performance of the active devices in the circuit, in particular the transistors. A bipolar circuit might, for example, have a 0.005 Ω-m epitaxial layer which is subjected to a limited source diffusion of acceptor impurities with a surface concentration of, say, 10^{24} m^{-3}, to a depth of a few microns, to form all base-type areas. This might be followed by a shallower constant source diffusion with surface donor concentration of 10^{26} m^{-3}, to form emitter-like regions. All passive planar components, such as diffused resistors or capacitors, have then to be designed so as to be fabricated from these layers, accepting the sheet resistances, doping levels, and profiles that exist in them.

In the following sections, the various types of circuit component that are available to the circuit designer will be discussed in detail.

5.1 Integrated circuit capacitors

Two basic types of capacitor are possible, each consisting essentially of two low resistance layers, or electrodes, separated by a dielectric. These are (a) thin-film capacitors in which the electrodes, one of which is usually metal, are separated by an insulating layer which is produced as an additional processing step after the planar diffusions and (b) diffused junction capacitors, which exploit the capacitance that exists between low resistance p and n regions, separated by a space-charge layer, which exists in a reverse biased junction.

In either case, the electrode areas, A, are usually large enough and their separation, d, small enough to neglect fringing fields, so the capacitance, C, is given by the usual expression:

$$C = \frac{\epsilon_r \epsilon_0 A}{d} \qquad\qquad (5.1)$$

or in practical terms, the capacitance per unit area is:

$$\frac{C}{A} = 8.85 \times 10^{-10} \frac{\epsilon_r}{d} \text{ pF } (\mu m)^{-2} \qquad\qquad (5.2)$$

5.1.1 Thin-film capacitors

Metal-oxide-silicon (MOS) capacitors can be formed using an emitter type diffusion as the bottom electrode, which has a low sheet resistance. This ensures that the series resistance included in the complete equivalent circuit of the component, to account for the parasitic resistance between its terminals and active region, called the *access resistance*, is kept to a minimum.

The top electrode, which is formed by the usual interconnection metallizing process, is separated from the other by a dielectric layer, as shown in Fig. 5.1(a). This can be either silicon dioxide, which has a relative permittivity of 3–4 or, less commonly, silicon nitride, with a relative permittivity which can approach 9. Both dielectrics have high dielectric strengths, to produce maximum breakdown voltages around 1 kV $(\mu m)^{-1}$, so that the available capacitance can be increased by thinning the layer by etching, as shown. This would, however, require additional processing steps. Typical oxide thicknesses are in the range 1–0.1 μm, the lower limit being set by difficulties with pinholes on the dielectric.

The equivalent circuit of an MOS capacitor is complex, but if leakage resistance is neglected, which is usually possible, a simplified circuit, shown in Fig. 5.1(b) is applicable. Note that capacitor terminal T_1 must always be biased positively with respect to the substrate, S, to prevent conduction to it via the junction diode, D.

Metal-oxide-metal capacitors can be fabricated entirely on top of a diffused integrated circuit slice, as shown in Fig. 5.2. This involves more processing steps with the usual concomitant disadvantages of economy and yield. These are offset to some extent by the advantages that the dielectric material can be chosen to suit a particular application, the stray capacitance is reduced, and the access resistance minimized. Large value capacitors can also be fabricated over existing circuits for efficient use of chip area. The oxides of aluminium, tantalum, titanium, and hafnium have been employed as dielectrics in such capacitors.

5.1.2 Junction capacitors

The various *p–n* junctions that exist in planar monolithic structures have an associated capacitance per unit area which can be exploited to produce an integrated

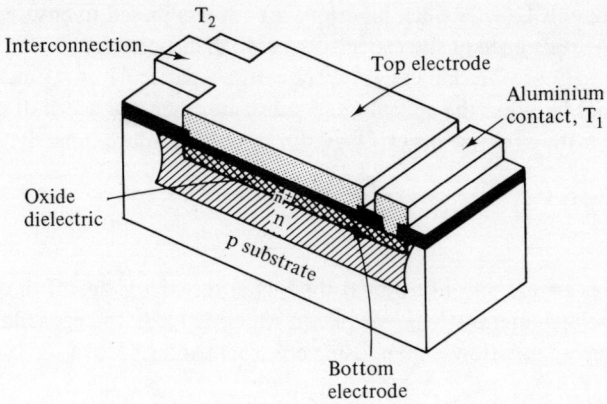

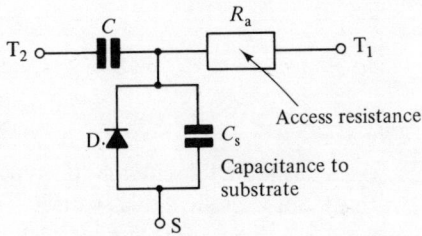

Fig. 5.1 Integrated MOS capacitor and its equivalent circuit.

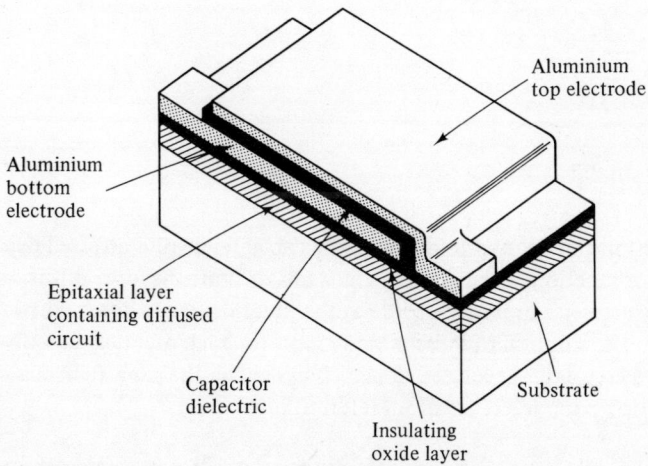

Fig. 5.2 Thin-film capacitor.

circuit component. Usually, such junctions are reverse biased to ensure that the effective shunt resistance of the capacitor is as high as possible.

If the impurity profile changes abruptly across a particular $p-n$ junction, as for example it does between the epilayer and substrate, then the width of the depletion layer, d_j, with a reverse bias voltage, V_r, applied is approximately:

$$d_j \simeq \left(\frac{2\epsilon_r \epsilon_0 V_r}{eN}\right)^{1/2}$$ (5.3)

where N is the carrier concentration at the higher resistivity side of the junction and ϵ_r is the relative permittivity for silicon which is 11.8. The capacitance per unit area of such a junction is then, using eqs. (5.1) and (5.3):

$$\frac{C}{A} = \left(\frac{e\epsilon_r \epsilon_0 N}{2V_r}\right)^{1/2}$$ (5.4)

or

$$\frac{C}{A} \propto (V_r)^{-1/2}$$

An obvious disadvantage which has to be tolerated in this type of component is that its capacitance is not fixed, but varies with bias voltage.

One or other of the diffused $p-n$ junctions in the epitaxial layer is often used to form the most commonly occurring integrated circuit capacitor. The impurity profile across such a junction depends on the type of diffusion used and varies gradually rather than abruptly across the junction plane. If the slope, s, of the impurity profile can be assumed constant, then the capacitance per unit area of the *linearly graded* junction is given by:

$$\frac{C}{A} = \left(\frac{\epsilon_r^2 \epsilon_0^2 es}{12V_j}\right)^{1/3}$$ (5.5)

or

$$\frac{C}{A} \propto (V_j)^{-1/3}$$

This expression can be used to estimate the capacitance of a diffused capacitor. However, for junctions formed using constant- or limited-source diffusions, the slope of the effective impurity profile at the junction is not linear, but varies with depth. The solution of Poisson's equation for such practical junctions is difficult and computer-generated design charts giving the capacitance as a function of the junction parameters are often referred to.*

* See, for example, Lawrence, H. and Warner, R. M. 'Diffused junction Depletion Layer Calculation', *Bell Syst. Tech. J.*, **39**, 389, 1960.

A typical integrated diffused capacitor, which utilizes the capacitance between collector and base-type diffusions, is shown in Fig. 5.3(a). The capacitance, C, of such a component is typically of order $100 \, \text{pF mm}^{-2}$ with reverse bias voltages of around 1 V. The n^+ emitter-type diffusion ensures that a low resistance ohmic contact exists between the p-type aluminium interconnection and the n-type base diffusion. A possible equivalent circuit is also shown, in Fig. 5.3(b). The access resistance to the capacitor, R_a, could be high due to the high resistance path from T_2 through the epi-layer. However, its value is often reduced by the inclusion of

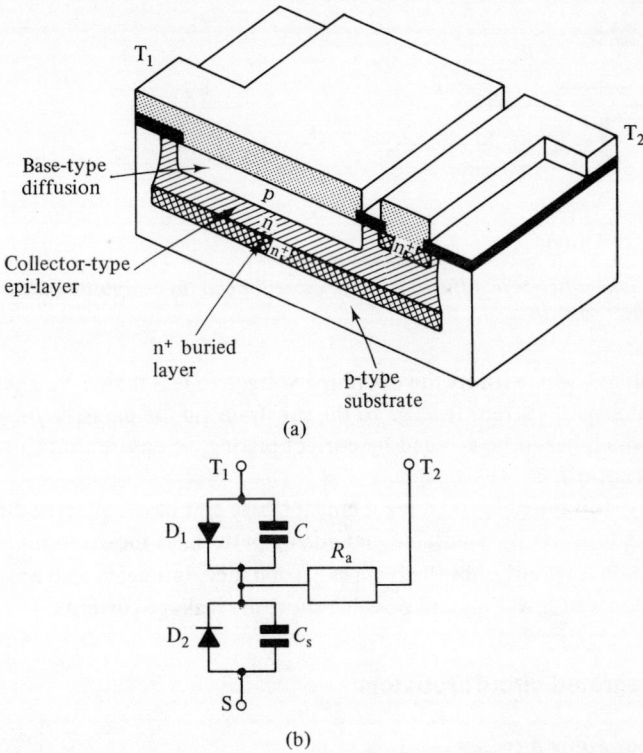

(a)

(b)

Fig. 5.3   A diffused junction capacitor and its equivalent circuit.

an n^+ buried layer diffusion, as shown. Junction diode D_1 is always reverse biased to reduce shunt resistance, so T_2 is held positive with respect to T_1 (and also with respect to the substrate, to reverse bias D_2 and maintain component isolation, as discussed in the next chapter).

The breakdown voltage, V_{BD}, of the diffused capacitor just described is that associated with the base-collector junction. An increase in capacitance per unit chip area can be achieved by employing the emitter-base type junction, as shown in Fig. 5.4(a). The capacitance of this junction is typically around $10^3 \, \text{pF mm}^{-2}$ for a 1 V reverse bias voltage, but this high value is at the expense of a lower break-

57

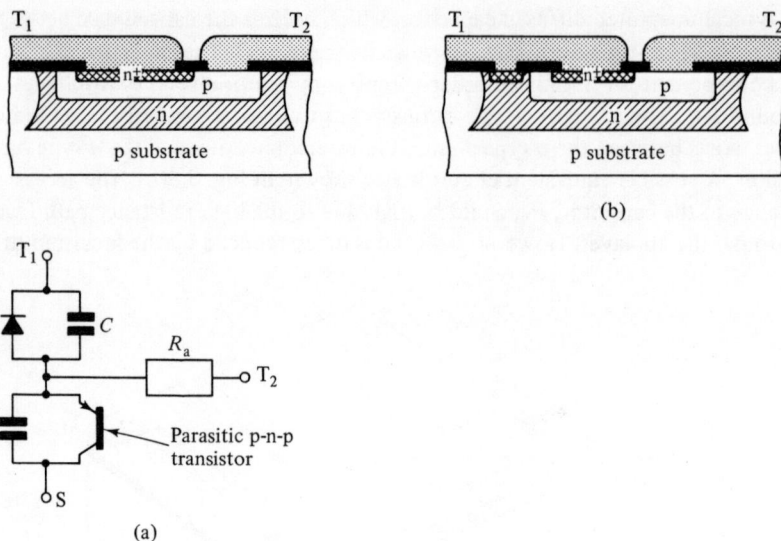

Fig. 5.4 (a) emitter/base diffused junction capacitor and (b) composite emitter/base and collector/base capacitor.

down voltage, which limits the operating voltage to less than 5 V. There is also the possibility of current leakage to the substrate via the parasitic p–n–p transistor shown, which has to be avoided by correct biasing, to ensure that this transistor is always cut-off.

Finally, it is possible to connect emitter/base and base/collector diffused capacitors together by a suitable metallizing pattern, as shown in Fig. 5.4(b). This combination obviously has the highest capacitance, but again also has the inherent disadvantages of low V_{BD} and possible substrate leakage currents.

5.2 Integrated circuit resistors

5.2.1 Monolithic diffused resistors

These are most conveniently formed by the same p-type diffusion as is used for transistor bases, into the usual isolating n-type island, as shown in Fig. 5.5. The junction between resistor and n-type island is always maintained in a reverse bias condition, to constrain currents to travel only through the resistor. The typical base diffusion used is a limited-source type, with a surface concentration of acceptors of around 10^{24} m^{-3} and a few micrometres depth, resulting in a sheet resistance, ρ_s, in the range 100–200 Ω/square. The resistance of the diffused resistor, R, is then determined by this figure for ρ_s and the ratio of the resistor length, l, to its width, w, since

$$R = \rho_s l/w \qquad (5.6)$$

58

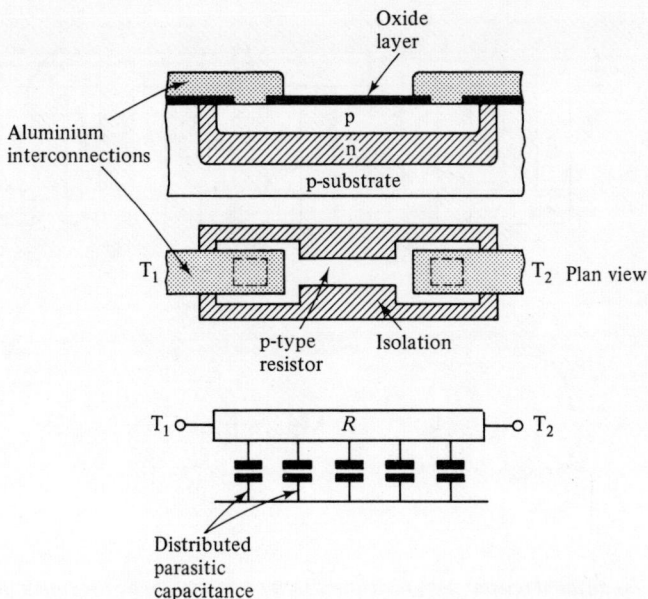

Fig. 5.5   Diffused integrated resistor and a possible equivalent circuit.

For example, if $\rho_s = 100\ \Omega$/square, a 1 kΩ resistor can be formed from ten squares in series, i.e., it is ten times as long as it is wide. Resistances in the range 50 Ω–10 kΩ, with a tolerance of around ±25 per cent, are obtainable with this technique.

High value resistors are often folded to conserve chip area. The turning points of such resistors can be metallized, as shown in Fig. 5.6(a), to avoid current crowding and provide low tolerance values, but this tends to be extravagant in area, so other geometries, as shown in Fig. 5.6(b) and (c) are mostly preferred. Since a drafting machine is used to cut master masks, as described earlier, resistor geometries are usually limited to straight lines, right angles, and arcs. Conformal mapping techniques are used to calculate the resistance of bends such as that shown in Fig. 5.6(b). For example, it can be shown that the resistance of the general bend shown in Fig. 5.7 is given by:

$$R_{\text{AB}-\text{CD}} = \rho_s \frac{\theta}{\log_e(r_2/r_1)} \qquad (5.7)$$

(see problem 5.8).

By these techniques, it can be shown that the effective resistances of the bends shown in Figs. 5.6(b) and 5.6(c), between planes A and B, are approximately equivalent to 3.9 and 3.8 squares respectively.

It is not always easy to make good electrical contact to the entire end strip width of a diffused resistor, so enlarged end contacts, as shown in Fig. 5.8, are

59

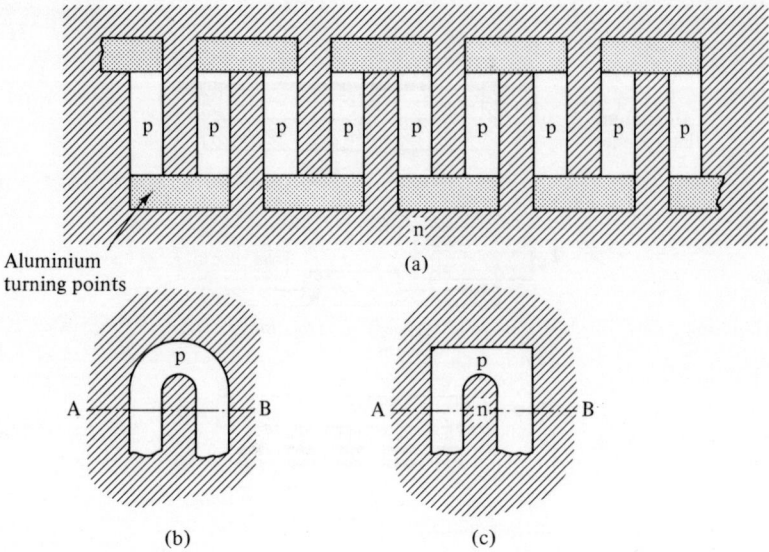

Fig. 5.6 *(a) diffused resistor with metallized turning points, (b) and (c) other possible turning-point arrangements.*

often formed. Such contacts can be easily made and produce a fairly uniform current flow; in the geometry shown they effectively extend the length of a diffused resistor by about 0.6 square.

Since the resistance of a diffused resistor is given by eq. (5.6), the choice of the width, w, effectively determines the surface area of a resistor of given resistance; the smaller w, the smaller the area. The lower limits for w are set by the power handling required, which is usually negligible for signal circuits, and the minimum tolerances that can be held using conventional masking and diffusion techniques. Line widths in the range 5–25 μm are usual; the larger widths can be held to better tolerances to produce more accurately reproducible resistors, but the narrower widths provide resistors with a smaller area and a better high-frequency performance. Narrower widths of around 1 μm are possible, with difficulty, the principal problems being associated with undercutting at the oxide window etching stage.

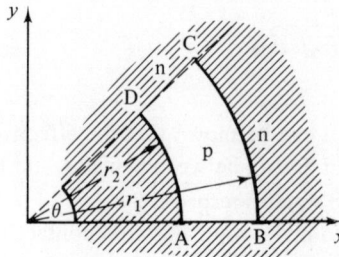

Fig. 5.7 *Generalized bend using circular arcs.*

60

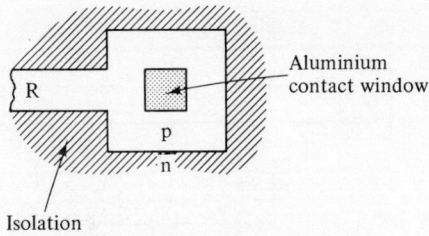

Fig. 5.8   End contact for a diffused resistor.

The equivalent circuit of a diffused resistor consists of a distributed resistance plus distributed stray capacitance to the substrate, as shown in Fig. 5.5. For many applications, this can be reduced to the simple lumped equivalent circuit shown in Fig. 5.9. The time constant for such a circuit, $RC/2$, can be used to predict roughly the high-frequency cut-off frequency of a particular diffused resistor.

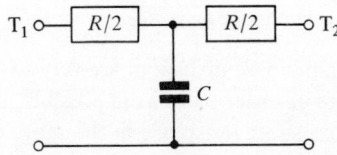

Fig. 5.9   A simplified equivalent circuit of a diffused resistor.

It will be further noticed from Fig. 5.5 that a parasitic *p–n–p* transistor exists between the resistor, isolation and substrate layers. The epitaxial layer is usually tied to the most positive point in the integrated circuit to ensure that this transistor is non-conducting.

5.2.2 Sheet resistance of diffused layers

It is not straightforward to calculate the sheet resistance of the diffused layers in which diffused resistors are formed. For example, the sheet resistance, ρ_s, of a thin slice, dt, of a diffusion of total depth t, assuming a uniform doping concentration of N in the slice, is

$$\rho_s = \frac{1}{\sigma \, dt} = \frac{1}{Ne\mu \, dt} \tag{5.8}$$

where μ is the appropriate carrier mobility. Since the layer is formed by a Gaussian diffusion, with density profile $N(t)$, say, the sheet resistance of the layer is then:

$$\rho_s = \frac{1}{e\mu \int_0^t N(t) \, dt} \; \Omega/\text{square} \tag{5.9}$$

61

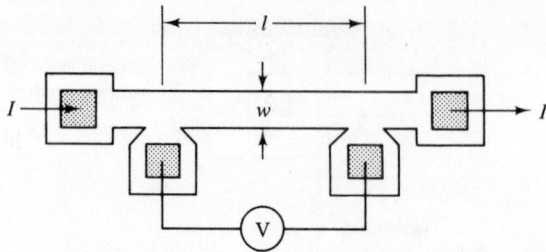

Fig. 5.10 A sheet-resistance test pattern for evaluating diffused layers.

Hence ρ_s depends on the type of diffusion, the junction depth, the average mobility, and the background and surface concentrations. Whereas it is evident that ρ_s can be estimated from such a calculation, it is usually monitored during initial diffusion evaluation tests, using, possibly, a sheet resistance test pattern, as shown in Fig. 5.10.

5.2.3 Low- and high-value diffused resistors

Low-value resistors, using the base diffusion, are very extravagant of chip area. However, resistance values of order of 1 Ω are possible, if an emitter type diffusion is used, which usually has a sheet resistance in the range 1–10 Ω/sq. A possible arrangement is illustrated in Fig. 5.11(a).

Higher than usual resistance values can be formed in the base region if its effective cross-section is reduced by an n^+ emitter diffusion, as shown in Fig.

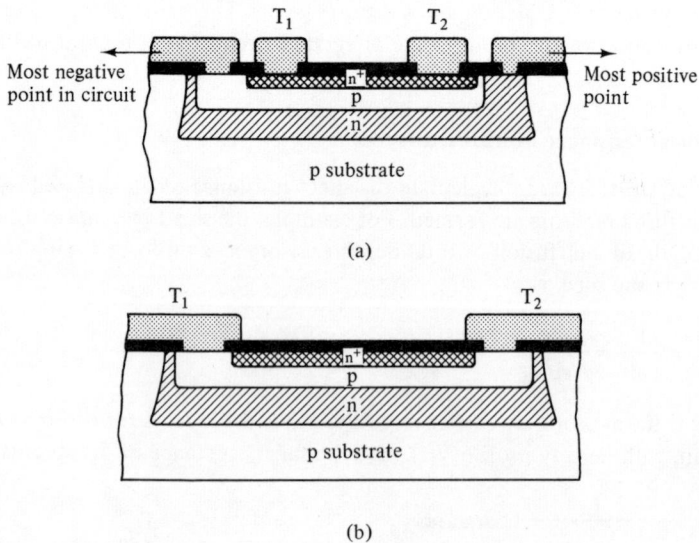

(a)

(b)

Fig. 5.11   Cross-sections of possible (a) low-value and (b) high-value diffused integrated resistors.

62

5.11(b). Current is in this way constrained to flow only in the p-region and sheet resistances of better than 1 kΩ/square become possible. One difficulty is that this structure bears some resemblance to a p-channel JFET, described later, and its resistance is a function of applied voltages.

The epitaxial layer itself can be used to produce high-value integrated resistors, using the isolation pattern to define their shape. These are not much used because of the relatively wide spread of tolerances involved.

5.2.4 Thin-film resistors

It is possible to deposit thin-film resistors on top of the diffused integrated circuit chip, by, for example, vacuum evaporation techniques, as shown in Fig. 5.12. This requires additional processing steps, but the associated disadvantages can sometimes be offset by a saving in total chip area per circuit. Further, there is a wide variety of resistance materials available so the sheet resistance can be chosen

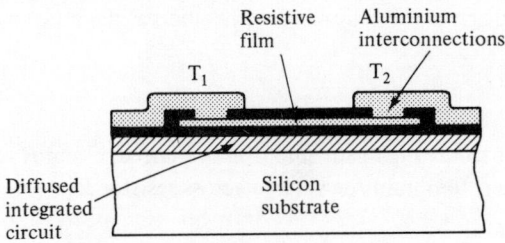

Fig. 5.12   Integrated thin-film resistor.

to suit a particular requirement. The lower associated stray capacitance also can improve the high-frequency performance of a resistor. Typical materials used for thin-film resistors are tantalum, tin oxide, and Nichrome, all of which have sheet resistances in the range 10–10^3 Ω/square.

5.3 Integrated circuit inductors

Whereas integrated inductors can be envisaged using, for example, a metallized spiral, they are best avoided because they are wasteful of chip space and because three-dimensional geometries are required to obtain high Q components.

Usually any necessary circuit inductance is provided by a lumped discrete component, for example an I.F. transformer, which is wired to the integrated circuit externally.

Alternatively it is possible to simulate an inductance, using more easily realizable integrated components, for example a passive resistor–capacitor combination, in conjunction with associated integrated active circuitry. Consider, as an example of the technique, the circuit shown in Fig. 5.13. The high-gain d.c. operational amplifier, G, and the other components can all be realized in integrated form, as

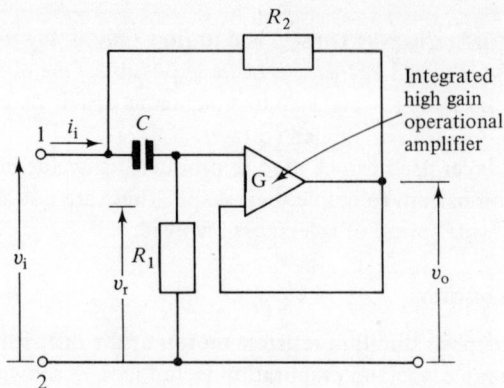

Fig. 5.13 *Possible integrated circuit for inductor simulation.*

described, for example, in section 7.11. The operational amplifier has its entire output voltage, v_o, fed back negatively to its inverting input, as shown. Summing currents at the input gives the following expression for the input current i_i

$$i_i = \frac{v_i - v_o}{R_2} + (v_i - v_r)j\omega C$$

If the input current to the high-gain amplifier is ignored, C and R_1 can be treated as a potential divider, to obtain the voltage across resistor R_1:

$$v_r = v_i\left(\frac{R}{R_1 - j/\omega C}\right)$$

These equations can be rearranged to obtain the input impedance, z_i, if $v_o = v_r$ is assumed, to give:

$$z_i = \frac{v_i}{i_i} = \frac{R_2(1 + R_1 R_2 \omega^2 C^2)}{1 + \omega^2 R_2^2 C^2} + j\frac{\omega R_2 C(R_1 - R_2)}{1 + \omega^2 R_2^2 C^2}$$

Therefore, provided $R_1 > R_2$, the circuit combination, when viewed from the input terminals, behaves as a frequency-dependent inductance. It can also be shown that if $R_1 \gg R_2$, the simulated inductance can have a high Q factor. For example, a circuit employing a 0.1 μF externally connected capacitor could function as a 1 H inductor with a Q value of 15 at 0.5 kHz, but such high capacitance values are not practicable using integrated capacitors, so inductance simulated using such components will be correspondingly reduced.

It is also possible to produce a simulated integrated tuned resonant circuit, using the arrangement of the type shown in Fig. 5.13, by providing an additional capacitor across the input terminals, which resonates with the simulated inductance. Obviously such procedures are costly in terms of chip area and yield and it is more usual for inductors and tuned circuits to be avoided if possible or added externally to the integrated circuit.

64

An alternative approach is to eliminate inductors and capacitors in tuned filter circuits altogether, by employing an analogue mechanical resonant circuit such as a surface or bulk acoustic wave in a piezo-electric material.

5.4 Bipolar integrated transistors

5.4.1 Monolithic integrated transistor

The cross-section and plan view of a typical general purpose, n–p–n, planar, monolithic bipolar transistor are shown in Fig. 5.14. As usual, the drawings are not precisely to scale, for clarity, but some possible dimensions are included to give an idea of the sizes involved. The transistor is entirely fabricated in an n-type epitaxial layer supported by a p-type substrate, as described earlier, and is electrically isolated from other components in the layer by junction isolation (see chapter 6). A p-type diffusion through the epi-layer to the substrate provides isolation and defines the extremities of the collector. This is followed successively by a limited-source p-type base diffusion and a constant-source n^+ emitter diffu-

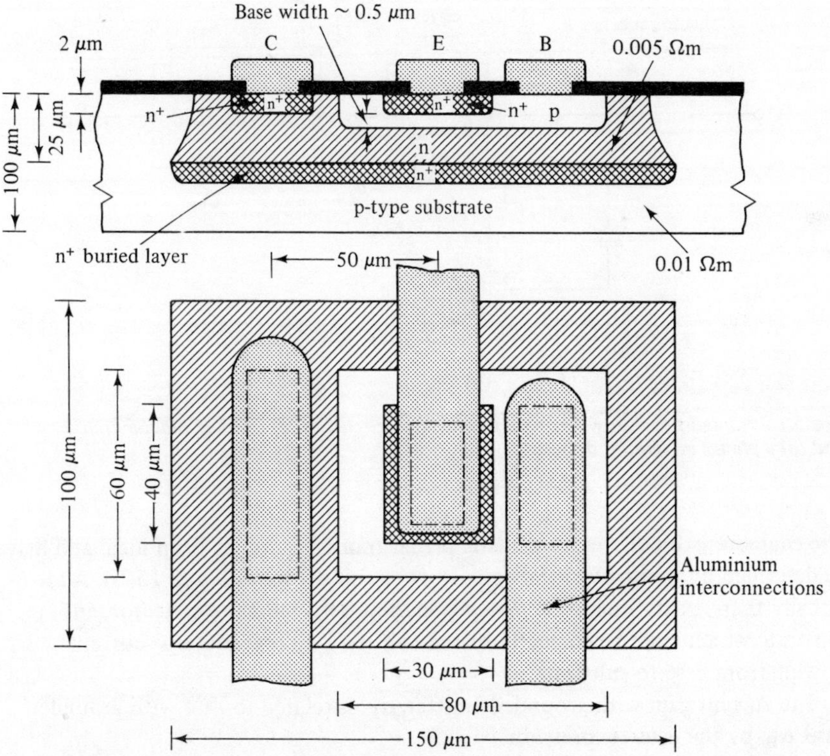

Fig. 5.14   General purpose, n-p-n planar, monolithic, integrated transistor.

65

sion, as shown in Fig. 4.14. This processing results in the *graded-base* doping profile illustrated in the figure.

The equivalent circuit of an ideal transistor would be as shown in Fig. 5.15(a) and its operation could be entirely defined by a current-gain, α_E, where

$$\alpha_E = h_{fE} = i_c/i_b \qquad (5.10)$$

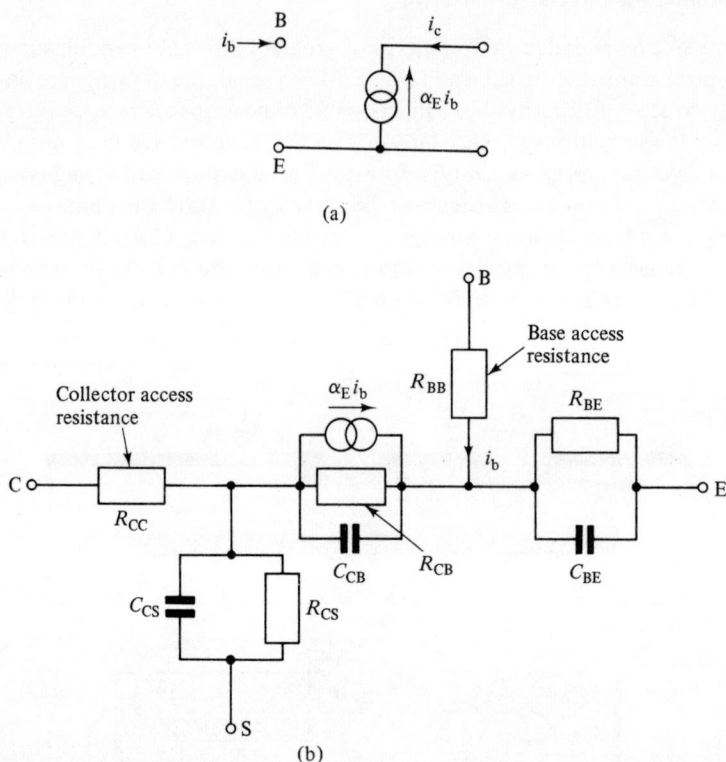

(a)

(b)

Fig. 5.15 Small-signal, low-frequency equivalent circuit of (a) an ideal bipolar transistor and (b) a planar integrated device.

The characteristics of a more realistic planar transistor are far from ideal and may be described by the small-signal equivalent circuit shown in Fig. 5.15(b). Additionally there exists a parasitic *p–n–p* transistor between base, collector, and substrate which must be correctly biased to cut-off to prevent large currents flowing from base to substrate.

The current gain with grounded emitter, α_E, is related to that with grounded base, α_B, by the usual expression

$$\alpha_E = \alpha_B/(1 - \alpha_B) \qquad (5.11)$$

The value of α_B, which should approach unity for high gains, is dependent on the product of four factors, as follows:

$$\alpha_B = \eta_E . \beta . \gamma_C . M \tag{5.12}$$

Each factor is influenced by device design and fabrication processes:

(a) Emitter injection efficiency, η_E

This is defined as the ratio of the electron current injected into the base of an n–p–n transistor to the total emitter-base junction current. It is given by:

$$\eta_E \simeq \left(1 + \frac{\sigma_B}{\sigma_E} . \frac{l_B}{l_E}\right)^{-1} \tag{5.13}$$

where the σ's are the conductivities and the l's the lengths of the base and emitter regions.* For efficiencies approaching unity, to produce high gain, the conductivity and length of the emitter region should both be much greater than the corresponding values for the base region; these conditions can be achieved by suitable diffusion programmes.

(b) Base transport factor, β

This is defined as the ratio of the electron current density at the collector junction of an n–p–n transistor, to that at the emitter junction, and is given by:

$$\beta = \frac{J_e|_{\text{collector}}}{J_e|_{\text{emitter}}} \simeq 1 - \tfrac{1}{2}\left(\frac{l_B}{L_e}\right)^2 \tag{5.14}$$

where l_B is the physical effective length of the base and L_e is the diffusion length of minority electrons in it.* It is evident, that for β to approach unity, the condition for high gain, the base should be as short as possible.

(c) Collector multiplication factor, γ_c

This factor is defined as the ratio of the total current to the electron current at the collector-base junction. It may be assumed almost equal to unity for most planar transistors.

(d) Multiplication factor, M

Electrons entering the collector of an n–p–n transistor experience large electric fields, E, due to the collector voltage, V_{CB}. This can lead to avalanche breakdown for $E \approx 1 \text{ MVm}^{-1}$, or typically for V_{CB} of a few tens of volts. At lower voltages than this, some electron-hole pair production is possible by ionizing collisions but the voltage may not necessarily be sufficient to sustain avalanche breakdown.

* See for example, Allison, J. *Electronic Engineering Materials and Devices*, McGraw-Hill.

The increased number of carriers results in an increase in gain by a factor, M, found empirically to be of the form:

$$M \simeq (1 - V_{CB}/V_{BD})^{-n} \tag{5.15}$$

where V_{BD} is the breakdown voltage. Typically $n \simeq 4$ for n-type silicon and 2 for p-type.

5.4.2 Collector access resistance

The electrical connections to planar integrated transistors are all made from the top face of the silicon slice, which leads to finite *access resistances* between the terminals and the effective active regions to which they connect. In particular, the collector access resistance, R_{cc}, must be minimized if the transistor performance is not to be degraded under high collector current flow conditions. Using the dimensions and resistivities shown in Fig. 5.14 as a simple example, it is evident that if the buried layer is ignored for the moment, the access resistance is approximately:

$$R_{cc} \simeq \frac{\rho l}{A} = \frac{0.005(50 \times 10^{-6})}{(25 - 2)10^{-6} \cdot 40 \times 10^{-6}} \simeq 270 \; \Omega$$

This access resistance can be considerably reduced by the introduction of the n^+ buried layer, shown in Fig. 5.14. Such a layer has a sheet resistance of around 10 Ω/square, which is in parallel to the high resistance path from the collector contact, and so reduces the effective access resistance to:

$$R_{cc} \simeq \frac{\rho_s \cdot l}{w} = \frac{10 \times 50}{40} \simeq 12 \; \Omega$$

The access resistance has been reduced, by the buried layer diffusion, by more than an order of magnitude. It can be lowered still further by the inclusion of a *collector wall*, in which the n^+ collector contact diffusion is driven-in and extended to join with the buried layer. This technique is particularly useful for high current transistors.

5.4.3 Base resistance and emission crowding

This effect, which is illustrated in Fig. 5.16, occurs because of the planar transistor's geometry and the finite resistance of its base region; it is most apparent for high current operation. An $i_b r$ voltage drop occurs between the base terminal B and any point on the emitter-base junction, the drop being greatest at points farthest removed from B. Now the base current is exponentially dependent on the *effective* forward bias voltage at any particular point, V_{EBeff}, which is dependent on the base access resistive drop, or

$$i_b \simeq I_0 \exp \left[eV_{EBeff}/kT \right] \tag{5.16}$$

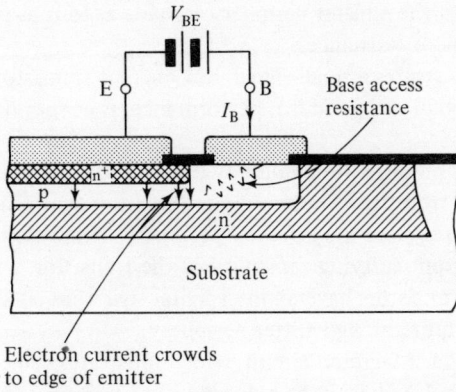

Fig. 5.16   *Emission crowding in a planar integrated transistor.*

It is evident that regions nearest to the base terminals have the least access resistance drop, hence the highest V_{EBeff}, and so pass most current, the rest of the junction remote from B passing hardly any. For modest currents, the non-uniform distribution of emitter current can be ignored but under high current conditions, the concentration of current at the periphery of the emitter, known as *current crowding*, can cause excessive current densities and damage there. Hence for high

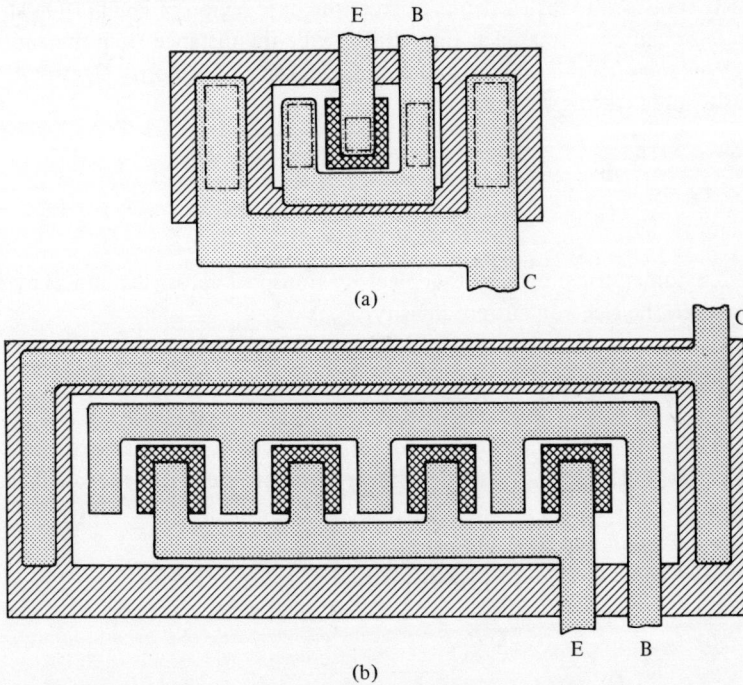

Fig. 5.17   *(a) Planar transistor with double base and collector contacts, (b), interdigital planar transistor.*

69

current transistors, the emitter periphery is made as long as possible, without introducing excessive resistance.

Figure 5.14 illustrates a small-signal, low current transistor that is simple to make, has a high gain and good h.f. performance, is economic in area and has a low noise capability.

If the design is modified to include two base and collector contacts, as shown in Fig. 5.17(a), current flows in two directions from the emitter so reducing the access resistance, R_{cc}, and the effective periphery of the emitter is doubled, increasing the current-carrying capability of the transistor.

High current transistors have an interdigital structure, as shown in Fig. 5.17(b), to produce the required long emitter periphery.

A planar, dipolar integrated circuit, which illustrates some of the components discussed so far, is shown in Plate 5.4.

5.4.4 High-frequency performance of transistors

The minority carriers take a finite time to traverse the base region of a transistor and it is this *transit-time*, τ_B, which is usually the major factor limiting high-frequency performance. As the operating frequency approaches τ_B^{-1} the transistor becomes inoperative.

The transit-time for electrons to cross the base region of length l_B in an *n-p-n* transistor can be estimated as follows. Let x be the distance from the emitter-base junction and $v(x)$ the velocity and $n(x)$ the density of minority electrons. The transit-time is then given by:

$$\tau_B = \int_0^{l_B} \frac{dx}{v(x)} \tag{5.17}$$

If it is assumed for simplicity that electron transport across the base is by diffusion only, then the electron current density, J_e, is

$$J_e = ev(x)n(x) = eD_e \frac{dn}{dx} \tag{5.18}$$

where D_e is the diffusion constant for minority electrons. An approximate solution can be obtained by assuming a linear gradient of electrons across the base, so that:

$$J_e = ev(x)n_0 \left(1 - \frac{x}{l_B}\right) = eD_e \left(-\frac{n_0}{l_B}\right)$$

or

$$v(x) = \frac{D_e}{l_B - x} \tag{5.19}$$

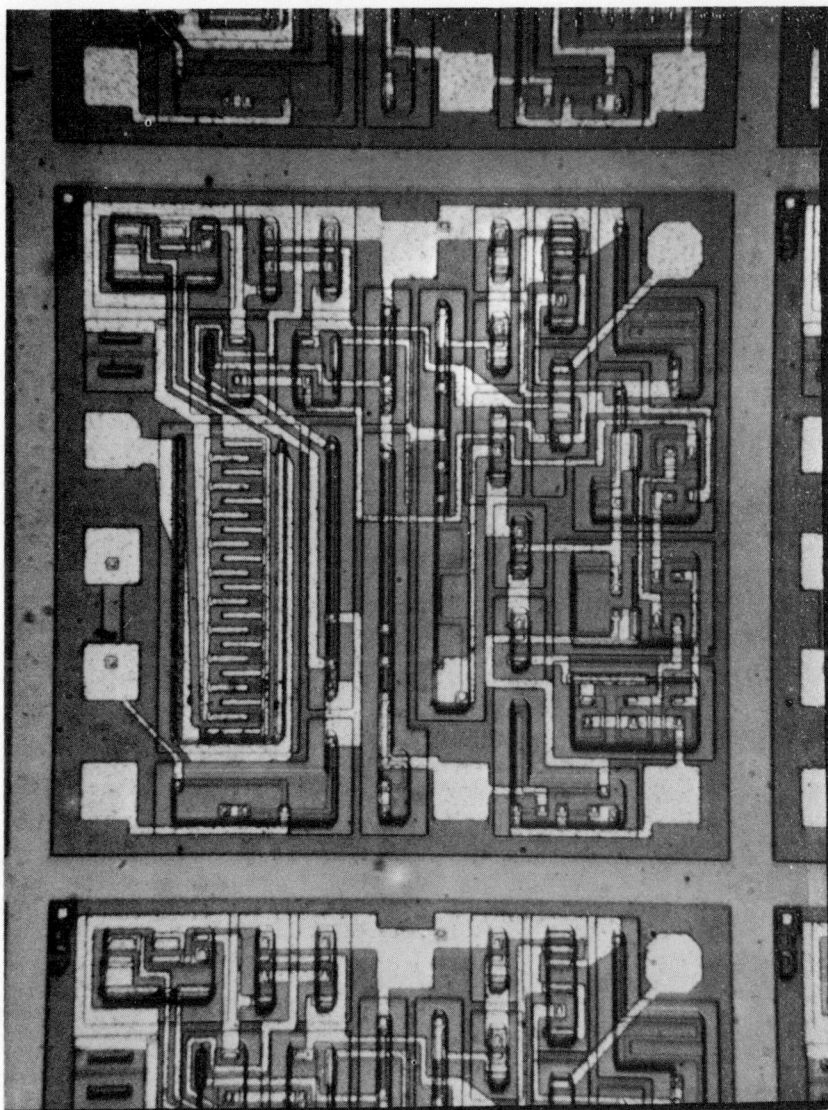

Plate 5.4   A microphotograph of a bipolar integrated alternator regulator circuit, which uses high-voltage techniques and is for automotive applications. Notice the folded diffused resistors, transistors, interconnections and bonding pads. Adjacent identical circuits on the silicon slice will also be seen. Reproduced by kind permission of Texas Instruments Ltd, Components Group.

Substituting this expression in eq. (5.17) and performing the integration gives:

$$\tau_B \simeq \frac{l_B{}^2}{2D_e}$$

(5.20)

It is evident that τ_B can be reduced, and the high-frequency performance improved, by making the effective length of the base, l_B, as thin as possible. A similar expression for the transit-time for holes in a p–n–p transistor could be derived, which would include D_h, the diffusion constant for holes. However since $D_e > D_h$, n–p–n transistors have a smaller transit-time than their p–n–p equivalents and are to be preferred for high-frequency or high-speed operation. The base transit-time can be further reduced if electrons are accelerated across the base by an electric field, rather than moving solely under the influence of diffusion. This property is exploited in *drift transistors*, which have a built-in drift field in the base region arising from the doping profile there.

There is an additional transit-time, τ_c, associated with carriers moving through the collector-base depletion layer. For collector voltages, V_{CB}, greater than a few volts, electrons in this region reach a saturated velocity, v_{sat}, of around 10^5 ms^{-1}. If the depletion layer thickness, which itself is dependent on the collector voltage, resistivity and so on, is d_{CB}, then:

$$\tau_c \simeq d_{CB}/v_s$$

(5.21)

Finally there are two more time constants, associated with the capacitance of the p–n junctions in a transistor. Firstly, the time constant due to the emitter-base junction, τ_{EB} is given by:

$$\tau_{EB} \simeq r_B \cdot C_{EB}$$

(5.22)

where C_{EB} is the capacitance of the forward biased junction and r_B its slope resistance. Substituting approximate values for these gives:

$$\tau_{EB} \simeq A \left(\frac{kT}{eI_e} \right) \left(\frac{e\epsilon_r\epsilon_0 N_{0B}}{2V_j} \right)^{1/2}$$

(5.23)

where N_{0B} is the acceptor concentration in the base near to the emitter and V_j the net junction voltage. Hence, in order to reduce τ_{EB}, the junction area, A, and N_{0B} must be kept as small as possible.

The time constant related to the base-collector junction, τ_{BC}, is given by

$$\tau_{BC} \simeq R_{cc}C_{CB}$$

(5.24)

which can be reduced by decreasing the collector access resistance, R_{cc}, or the doping concentration in the collector, or by increasing V_{CB}.

72

The sum of all four time constants provides a value for the total delay time, τ_t. The high-frequency operating limit corresponding to a particular value of τ_t, f_α, is then approximately:

$$f_\alpha \simeq (\tau_t)^{-1} \tag{5.25}$$

Here, f_α is the *alpha cut-off frequency*, at which the gain, α_B, falls 3 dB below its low-frequency value.

5.5 Integrated diodes

5.5.1 Switching characteristics

Since integrated circuit diodes are most often used in a switching mode, a preliminary discussion of this form of operation will demonstrate which parameters must be optimized, and also forms the basis for the comparison of different diode configurations.

Consider, for example, the forward biased n^+-p junction diode shown in Fig. 5.18. The minority electron concentration in the p-region will be approximately linear, as shown, if the base is short compared to the diffusion length, L_e. If the junction is suddenly reverse biased, at t_0, then, because of this stored electronic charge, the reverse current, I_R, is initially of the same magnitude as the forward current, I_F. However, as the stored electrons are removed into the n^+ region and the contact, the available charge quickly drops to an equilibrium level and a steady current eventually flows corresponding to the reverse bias voltage, as shown in Fig. 5.18.

To estimate the *turn-off time*, t_{off}, let us first calculate the minority charge stored at the junction when a forward current, I_F, flows. The velocity of electrons distance x from the junction, v, is given by eq. (5.19), so the charge density at x, $\rho(x)$, is

$$\rho(x) = n(x)e = \frac{I_F}{Av} = \frac{I_F}{A} \cdot \frac{l_p - x}{D_e} \tag{5.26}$$

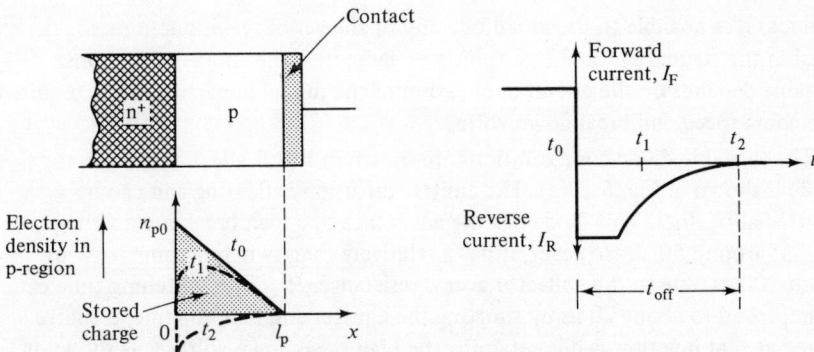

Fig. 5.18 Charge stored and current flow in a switched n$^+$-p *junction diode.*

73

The total stored charge stored, Q, is then:

$$Q = A \int_0^{l_p} \rho(x) \, dx = \frac{I_F l_p^2}{2D_e} \qquad (5.27)$$

When the voltage is reversed, the average reverse current flowing during the turn-off period, $\bar{I}_R$ is given by:

$$\bar{I}_R = Q/t_{off}$$

or

$$t_{off} = \frac{Q}{\bar{I}_R} = \frac{I_F}{\bar{I}_R} \cdot \frac{l_p^2}{2D_e} \qquad (5.28)$$

Alternatively, this can be written in terms of the transit-time, τ_B, using eq. (5.20), to give

$$\boxed{t_{off} = \frac{I_F}{\bar{I}_R} \cdot \tau_B} \qquad (5.29)$$

Thus the turn-off time, because of its dependence on I_F and $\bar{I}_R$, is a function of the external circuitry to some extent, but it is also determined by a characteristic transit-time of the diode, τ_B, which can be reduced, for example, by shortening the length of the p-region.

The stored charge and consequently the switching time can be greatly reduced by the introduction of gold impurities into the junction diode, by diffusion. The gold dopant, sometimes called a *lifetime killer*, provides a series of recombination centres, at which stored minority carriers are removed more quickly because of the increased recombination rate. This technique is often used to produce diodes and other active devices for high-speed applications.

5.5.2 Integrated planar diode configurations

Whereas it is possible to use as a diode any of the various p–n junctions which exist in the standard monolithic transistor geometry, the choice of the most suitable depends on the circuit application of the diode, in particular the required switching speed and breakdown voltage.

The available diode configurations are shown in Fig. 5.19. A base-collector diode is shown in Fig. 5.19(a). The emitter diffusion is floating and can be omitted altogether. This diode has the advantage of a high breakdown voltage, V_{BD} of around 50 V. However, it has a relatively long switching time, t_{off}, of about 100 ns, due to the collector access resistance, R_{cc}. The switching time can be improved to about 70 ns by shorting the emitter and base to remove charge stored at that junction, while retaining the high breakdown voltage, as shown in Fig. 5.19(b).

74

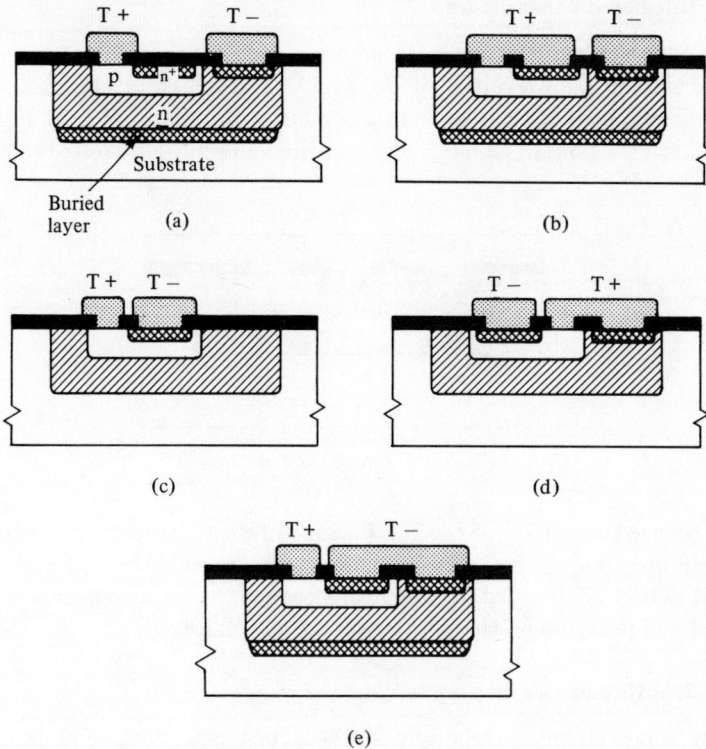

Fig. 5.19   Possible integrated monolithic diode configurations.

The base-emitter junction diode with collector floating is illustrated in Fig. 5.19(c). The turn-off time, due mainly to charge stored in the base-collector junction, is about 80 ns and it has a low V_{BD}, associated with the highly doped emitter, of around 5 V. Again, the switching time can be reduced to as low as 20 ns, by shorting base and collector, to remove stored junction charge, as shown in Fig. 5.19(d), the low V_{BD} being unaffected.

Finally, it is possible to form a diode from the emitter-base and base-collector junctions in parallel, by shorting emitter and collector, as shown in Fig. 5.19(e). However, this arrangement is little used because of its high associated junction capacitance, which produces a low switching speed of around 150 ns, together with a poor breakdown voltage of about 5 V, associated with the base-emitter junction.

Obviously, the most widely used diode geometries are those illustrated in Figs. 5.19(b) and (d), the former for higher voltage applications, and the latter where switching speed is of paramount importance. Both diodes are also used in less demanding circuits because of their opposite polarities, and sometimes a particular configuration is chosen so that the diode may be included in the same isolation region as a transistor with which it might be associated.

75

5.5.3 Integrated Zener diodes

It was noted in the previous section that the emitter-base junction has a low reverse bias breakdown voltage of around 5 V. This is used to advantage, to produce a Zener diode for voltage control circuits, as shown in Fig. 5.20. Emitter and collector regions of an integrated transistor structure are shorted together

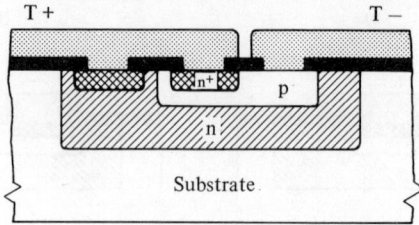

Fig. 5.20   Integrated Zener diode.

and a reverse bias voltage, exceeding V_{BD} and of the polarity shown, is applied between them and the base. The reference voltage provided by such a diode cannot usually be specified to close tolerances and the associated circuits must be designed to accommodate possible spreads in voltage.

5.5.4 Schottky barrier diodes

Carrier storage effects are relatively very small and usually can be neglected in the Schottky barrier diode, which is formed from the junction between certain metals and semiconductors. Such diodes are therefore extremely valuable components in fast switching circuits. In the silicon monolithic system, a Schottky barrier can

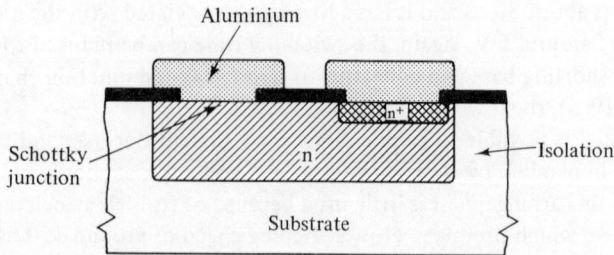

Fig. 5.21   Integrated Schottky barrier diode.

be formed between the epitaxial layer and the aluminium deposited for interconnections. A typical arrangement is shown in Fig. 5.21. The cathode connection to the epitaxial layer is via a conventional n^+ collector contact diffusion, to ensure a good ohmic contact, but at the anode connection the n^+ diffusion is omitted and a Schottky barrier is formed between aluminium and n-type epitaxial silicon.

76

5.6 Integrated field-effect devices

5.6.1 Junction field-effect transistors (JFETs)

Monolithic integrated JFETs can be conveniently manufactured using the conventional planar techniques already discussed. A typical n-channel, junction isolated, planar JFET is illustrated in Fig. 5.22. The n-type channel, formed in

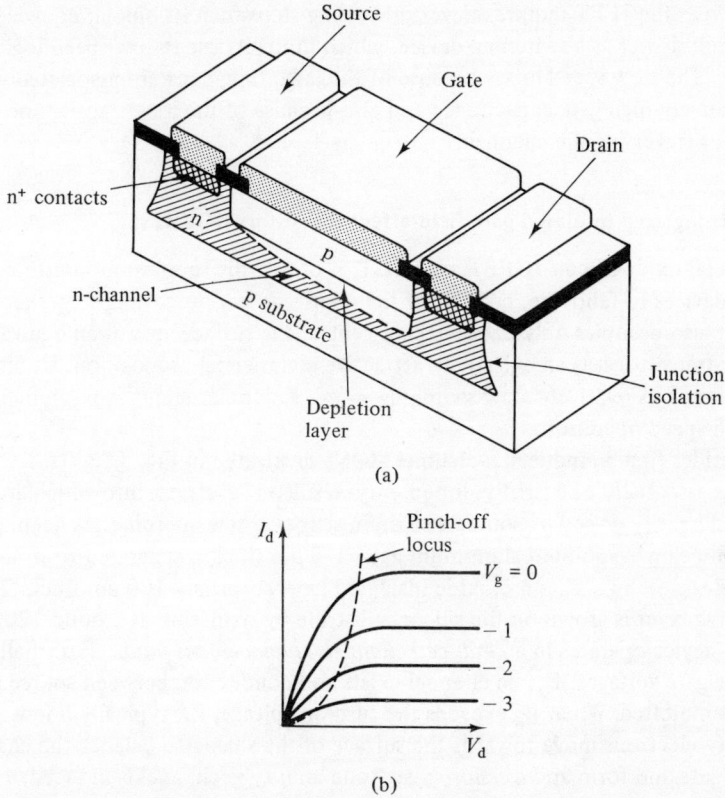

(a)

(b)

Fig. 5.22   (a) Integrated, planar, n-channel JFET, and (b) its drain characteristic.

the epitaxial layer, links source and drain and a p-type base-like diffusion acts as a gate. Note the symmetry of the geometry, which is often present, source and drain contacts being interchangeable.

The JFET is operated in the *depletion* mode. Since the p-type gate is more heavily doped than the n-type channel, the depletion layer which exists when the junction is reverse biased occurs mostly in the channel material. The total effective reverse bias voltage at the junction is the sum of two components, the gate-source voltage, V_g, and the voltage which exists in the conducting channel due to

77

the IR drop in it, which is dependent on the drain voltage, V_d, and is also a function of position. For a particular gate voltage, as the drain voltage increases, the total effective reverse bias and hence the depletion layer width increases. This causes the effective channel cross-sectional area to be reduced; its resistance increases and the rate of increase of drain-current is reduced. At some critical value of V_d, the depletion layer completely penetrates the channel and the drain current saturates. The drain characteristics, I_d vs. V_d, then have the general form shown in Fig. 5.22(b).

Whereas the JFET requires fewer processing steps than its bipolar equivalent, it is much slower as a switching device, which limits its use to low-speed logic circuits. The slow speed arises because of the large time constant associated with the relatively high gate capacitance and also because of the long transit time for electrons traversing the channel.

5.6.2 Integrated insulated-gate field-effect transistors (IGFETs)

The metal-oxide-silicon IGFET, or MOST, is one of the simplest monolithic active devices to fabricate, because of the relatively few processing stages required. Since it also occupies only about 10 per cent of the surface area of an equivalent bipolar transistor it is an extremely attractive commercial proposition. Its only limitation is its relatively slow switching speed, so bipolar circuits are preferable for high-speed operations.

Consider first an induced n-channel MOST illustrated in Fig. 5.23. This consists essentially of a lightly doped, p-type silicon substrate, into which are diffused heavily doped n^+ source and drain stripes, a few micrometres deep. An electron-beam evaporated aluminium gate, 1–2 μm thick is separated from the channel region by a silicon dioxide insulated layer, typically 100 nm thick. The insulating layer is grown on the silicon substrate by oxidation at around 1200°C.

The device operates in an *induced channel, enhancement mode*. For small positive gate voltages, V_g, no channel exists and conduction between source and drain is inhibited. When V_g exceeds the turn-on voltage, V_t, typically a few volts, minority electrons move towards the surface of the silicon to balance the charge on the gate and form an *inversion layer** and an n-type channel is induced. Conduction can then occur between source and drain if a suitable voltage, V_d, is applied. As V_g increases further, the resistance of the induced channel drops as more electrons are introduced into it, and I_d, the drain current, increases. Typical drain curves are shown in Fig. 5.23(b). This class of device is called *normally off*, since, for $V_g = 0$, no channel exists and $I_d = 0$.

Since during its operation all the junctions are reverse biased, the p-type substrate being connected to the most negative point in the circuit, the integrated MOST possesses the most important property of being self-isolating. This elimin-

---

* For more physical details of this process refer, for example, to Allison, J., *Electronic Engineering Materials and Devices*, McGraw-Hill.

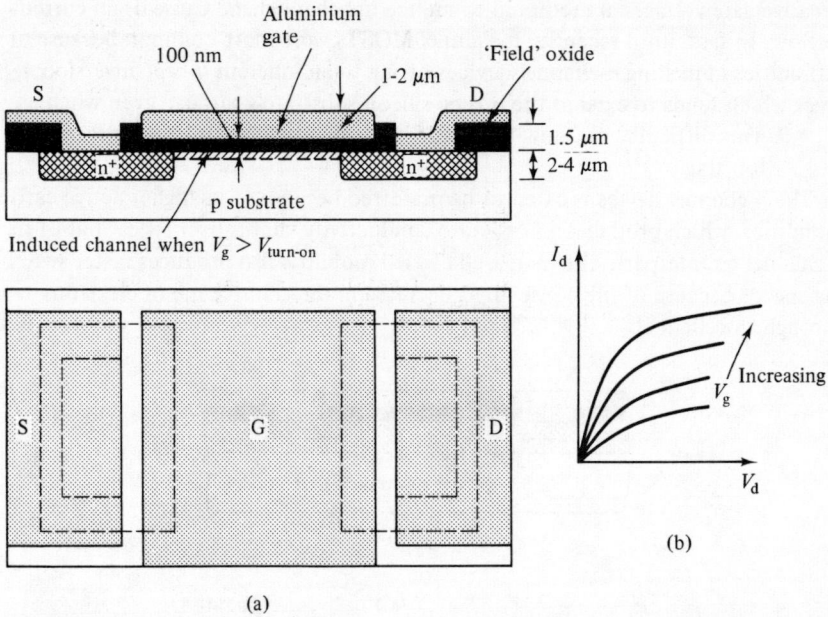

Fig. 5.23 *(a) Induced n-channel, enhancement, integrated MOST and (b) its drain characteristics.*

ates the need for special isolation diffusions, as for bipolar transistors, which results in a small area device and potentially large packing densities.

One difficulty to be avoided with this device is that it is possible for one of the interconnections, when at a high positive voltage, to behave as the gate of a parasitic MOST and cause conduction between n^+ diffusions over which it crosses. This can be avoided by effectively increasing the turn-on voltage outside the channel regions, by increasing the oxide thickness, as shown in Fig. 5.23. This thicker *field oxide*, which is typically 1 μm thick or more, is often deposited by the thermal oxidation of silane (SH_4) at about 400°C.

Sometimes it is necessary to reduce the turn-on voltage of an integrated MOST. This could be achieved by reducing the oxide thickness under the gate still further, to less than 100 nm, but difficulties are then often encountered due to pinholes. Alternatively, V_t can be reduced by increasing the relative permittivity of the gate insulation, but keeping its thickness constant. Silicon nitride (Si_3N_4), which has a permittivity about twice that of the oxide, is a most effective gate insulator, which can halve the turn-on voltage to 1–3 V. Since the silicon nitride-silicon interface is difficult to control, a thin oxide layer, about 30 nm thick, is first deposited, to separate effectively the nitride from the silicon in a metal-nitride-oxide–silicon or MNOS transistor. The reduced turn-on voltages in such MNOS devices result in a requirement for lower supply voltages.

An induced p-channel, normally-off device can be made with similar geometry to that shown in Fig. 5.23, except that n and p regions are interchanged and negative gate voltages are required to induce the channel and cause drain current to flow. In fact, until recently, p-channel MOSTs were more common because of difficulties in making n-channel devices owing to an inherent n-type inversion layer which tends to exist at the p-type silicon substrate's surface, even when $V_g = 0$. This difficulty can now be avoided by using a silicon nitride surface passivation stage.

The n-channel device is often to be preferred because of its higher active carrier mobilities, which produces an ON-state conductivity that is over twice that of its p-channel counterpart. The increased carried mobility also produces faster switching speeds because of the consequent decrease in the transit time of electrons through the channel.

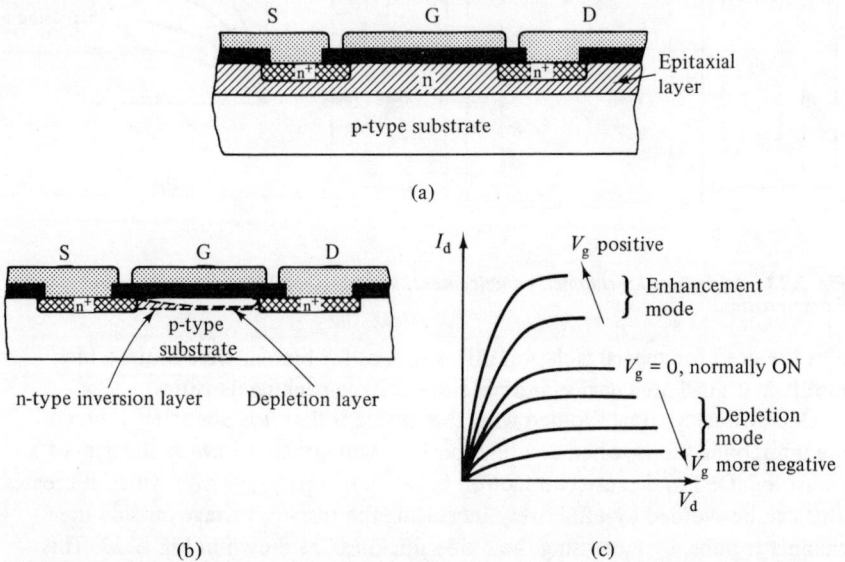

Fig. 5.24   (a) A metallurgical-channel MOST, (b) an inversion-layer MOST, and (c) the typical form of their drain characteristics.

Another variant of the MOS technology results in *normally-on* devices, typified by the *metallurgical-channel* MOST, shown in Fig. 5.24(a). In this device, heavily doped n^+ source and drain regions are diffused into an n-type epitaxial layer which acts as the permanent channel. Conduction between source and drain is therefore possible, even for zero gate voltages. Such MOSTs can be operated in either enhancement or depletion mode, depending on the sign of the gate voltage, V_g. If V_g is positive, electrons move into the channel region, to balance the charge on the gate, the channel conductance and hence I_d, is increased. For negative gate voltages, a depletion layer is formed under the gate electrode, to provide the necessary positive charge; this reduces the channel cross-section, consequently decreasing its conductivity and I_d falls. The resulting drain characteristics for this versatile device are shown in Fig. 5.24(b). It should be noted that the JFET cannot operate in an enhancement mode since this would require the gate-channel junction to be forward biased and large gate currents would be drawn. This biasing arrangement is only possible in an IGFET because of the insulated gate which prohibits large gate currents.

Another type of normally-on MOST relies on the presence of an n-type inversion layer at the surface of a suitably prepared p-type silicon slice, because of the presence of donor-like states at the interface between the silicon and grown oxide. The inversion layer behaves like an induced channel even when $V_g = 0$, as shown in Fig. 5.24(c). Although normally operated in the depletion mode the inversion MOST can again be used in an enhancement mode, with drain characteristics of the type shown in Fig. 5.24(b), which allows large signal voltage excursions on the gate.

5.7 Ion implanted MOS transistors

In the MOS devices discussed so far, a certain amount of overlap, typically 5 μm, exists between the gate and source and drain diffusions, which is necessary because of registration tolerances in the masking process and lateral diffusion. Such overlap tends to increase the capacitance of the gate and so reduce switching speed. Ion implantation fabrication technologies have been developed to reduce overlap capacitance. In these processes, the source and drain are only partially diffused initially and the gate metallization is deposited to the required width, as shown for an induced p-channel device in Fig. 5.25(a). The entire slice is then flooded with boron ions which have been accelerated to a high velocity. Ions striking metallized regions are collected but those impinging directly on to the oxide

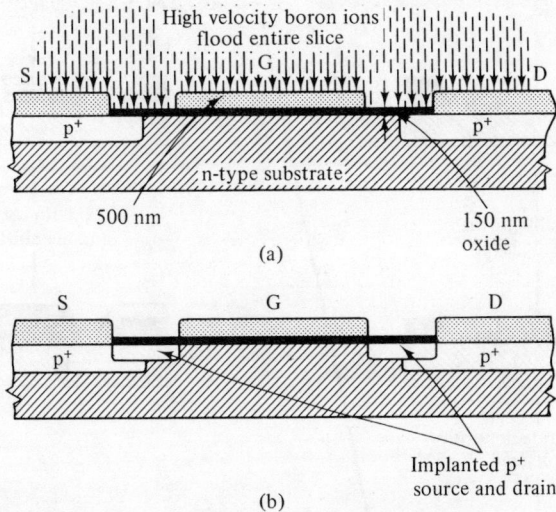

Fig. 5.25 Fabrication of an induced p-channel ion implanted MOS transistor. (a) Stage 1 and (b) finished device.

81

penetrate into the bulk silicon, to produce p^+ regions in the source and drain areas, as shown in Fig. 5.25(b). Notice that the process is self aligning since the gate metallization serves as an effective mask for the source and drain implantations. Consequently, the gate width and capacitance can be reduced, resulting in transistors which, typically, are 5x faster than an equivalent diffused device.

5.8 Silicon gate MOS transistors

The unique process used to produce this type of IGFET is that heavily doped amorphous silicon replaces the usual aluminium gate metallization to produce a deposited silicon gate electrode. The use of doped silicon rather than aluminium gate electrodes provides flexibility in the designed threshold voltage, which is dependent on the difference in work function between gate conductor and substrate semiconductor.

The basic fabrication steps to produce an induced p-channel device are outlined in Fig. 5.26. Once again the source and drain diffusions are self aligning because

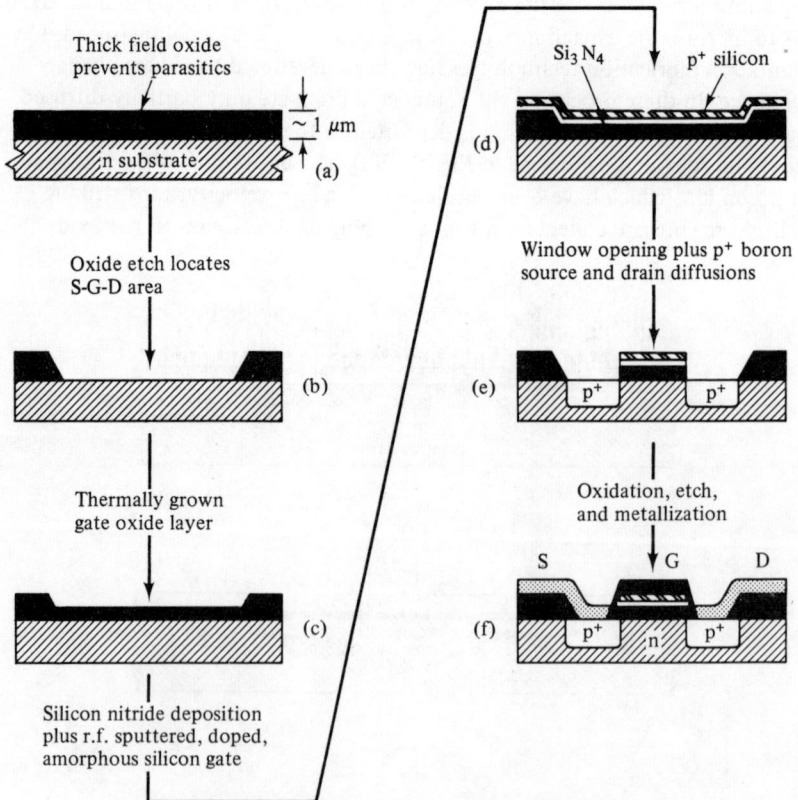

Fig. 5.26   Fabrication steps for an induced p-channel silicon gate IGFET.

of the presence of the silicon gate electrode which defines their extremities. Overlap is considerably reduced, although some residual overlap is caused by lateral diffusion, which increases the switching speed. A further advantage is an increased yield due to the thin gate oxides being immediately protected by the deposited silicon gate.

5.9 Other field-effect components

Whereas capacitors can be made which are suitable for MOS integrated circuits, as described in section 5.1, diffused resistors, as described in section 5.2, are not compatible with MOS technology. However, an MOS compatible field effect resistor can be made which utilizes the effective resistance of a MOST. The gate voltage of a MOST is held at a constant bias voltage of such a level that operation is in the nearly flat saturated part of the drain characteristic; see, for example, Fig. 5.24(b). This resistance between source and drain, which has a high dynamic value, can be used, say, as an effective load resistor for an active MOST.

It is often found more convenient in practice to use an IGFET as a resistor by permanently connecting its gate and drain together, using the metallized interconnection layer. This arrangement produces resistances typically in the kilo ohm range.

A field-effect current limiter can also be devised, in which the gate of an FET is permanently strapped to the source, so that, for drain voltages greater than the pinch-off voltage, the drain current saturates and remains substantially constant.

5.10 Hybrid integrated circuit technologies

5.10.1 Mixed discrete component bipolar/MOS circuits

We have seen that although the technologies used to produce junction bipolar transistors and field-effect transistors are similar, they are, in their standard versions, by no means compatible. For example, the isolation requirements of each class of device are quite different. Since compatible passive devices are available for circuits based on either class of device, it is unusual for general purpose integrated circuits to contain bipolar and field-effect components on the same chip.

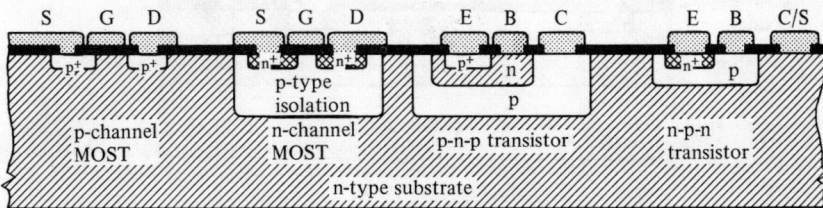

Fig. 5.27   Hybrid bipolar/MOS transistor technology.

83

In other words general circuits usually comprise either junction transistors plus diffused resistors and capacitors or IGFETs plus MOS resistors and capacitors.

Mixed technologies have been developed recently, however, and used successfully for more specialized circuits. For example, it has been found possible to accommodate complementary MOS transistor pairs (discussed later) and bipolar transistors plus related passive devices on the same circuit chip, as shown in Fig. 5.27. It will be noted that the isolation requirements of the various components can be successfully accommodated with this hybrid technology, albeit at the expense of additional processing steps, with the added difficulties of reduced yield and increased cost. However, such techniques can combine the advantages of high packing density and low static power dissipation of MOS components with the high speed capabilities of bipolars.

5.10.2 Bipolar insulated gate transistor (BIGFET) circuits

We have seen that IGFET circuits have distinct advantages of high yield due to their relative simplicity, low power dissipation, due to a high channel impedance, high input impedance and high packing density, but suffer from a low switching speed or frequency response which is to some extent a consequence of their high output impedance in conjunction with the capacitance of succeeding circuits. This difficulty can be overcome by combining discrete MOS and bipolar devices on the same current chip, as described in the previous section, the MOSTs carrying out logic processing and the bipolar transistors, which have a low output impedance, providing a high current output. Such discrete hybrid circuits introduce an additional complication, however, that of impedance matching between IGFET and bipolar devices. One possible solution is to taper the channel impedance of successive IGFET stages gradually by increasing the gate width and hence the channel conductance to match a final bipolar output stage.

Another more elegant solution, which has enjoyed some success, is to combine an IGFET and a bipolar transistor into one monolithic integrated output stage,

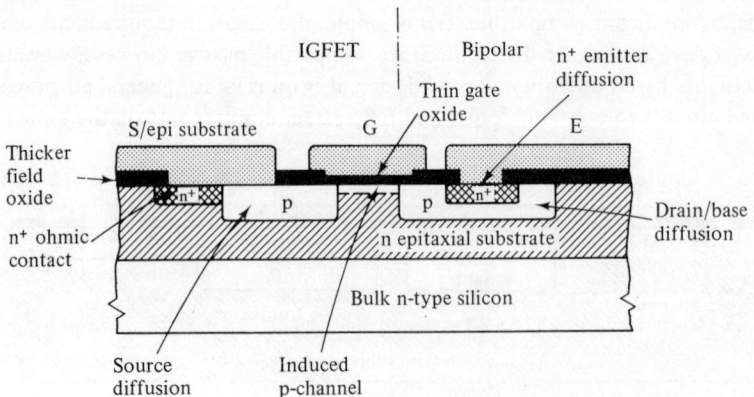

Fig. 5.28   Cross-section of a bipolar insulated gate transistor, or BIGFET.

the so called BIGFET device. A cross-section of this hybrid device is shown in Fig. 5.28. The base of the bipolar output transistor doubles as the drain for the preceding induced p-channel MOST and its collector is also common to the n-type epitaxial substrate of the MOST. Two identical n^+ diffusions provide the bipolar emitter and the usual ohmic connection to the epitaxial collector/substrate. The thin insulator below the gate is often made from an SiO_2–Al_2O_3 sandwich, which lowers the turn-on voltage of the MOST to around -1 V, so that the MOST can be operated from the same supply rail (-5 V) as the bipolar.

When the IGFET is turned on by a suitable negative voltage applied to its gate, a relatively small drain current flows, which constitutes a base current to the bipolar. The output emitter current is then this small base current, multiplied by the current gain of the bipolar, which is typically of order 100.

As a result of these properties, it is now not unusual for a BIGFET output stage to be incorporated in MOS logic circuits to allow direct interface with bipolar logic circuits, such as TTL, discussed later, with which they are compatible.

5.11 Collector diffused isolation (CDI) process

This is one example illustrative of several new technologies which have been developed recently to produce bipolar circuits, with their advantages of speed, drive-power, linear operation, and so on, but which require fewer processing steps and possess a much higher packing density than those employing conventional bipolar devices. In these respects CDI transistors are comparable to their MOS counterpart.

The basic processing stages used to produce a CDI integrated transistor are shown in Fig. 5.29. A p-type silicon substrate has n^+ buried layers diffused into it which eventually form the collectors of the bipolar transistors, Fig. 5.29(a). A high resistivity p-type epitaxial layer is next grown over the entire slice. The layer is made only a few micrometres thick, which is much thinner than in the earlier bipolar processes, so as to inhibit sideways diffusion by the n^+ isolation diffusion which follows, Fig. 5.29(b) and (c). Thus the device area, as defined by the extremities of the isolation diffusion is kept to a minimum. The n^+ diffusions also serve as high conductivity contacts to the buried collectors. A non-selective, shallow, p-type layer, with a resistivity which is suitable for diffused resistors and transistor bases, is then diffused over the entire slice, Fig. 5.29(d). A final shallow n^+ diffusion locates the emitters, Fig. 5.29(e). The previous shallow p-type diffusions are pushed in by this process, as shown, to form composite graded-base regions.

Some of the principal advantages claimed for the process are as follows:

(a) The processing steps are well established, and fewer are required than for conventional bipolar circuits. There are, typically, five masking operations and a saving of around eight additional stages, which naturally leads to an increased yield.

85

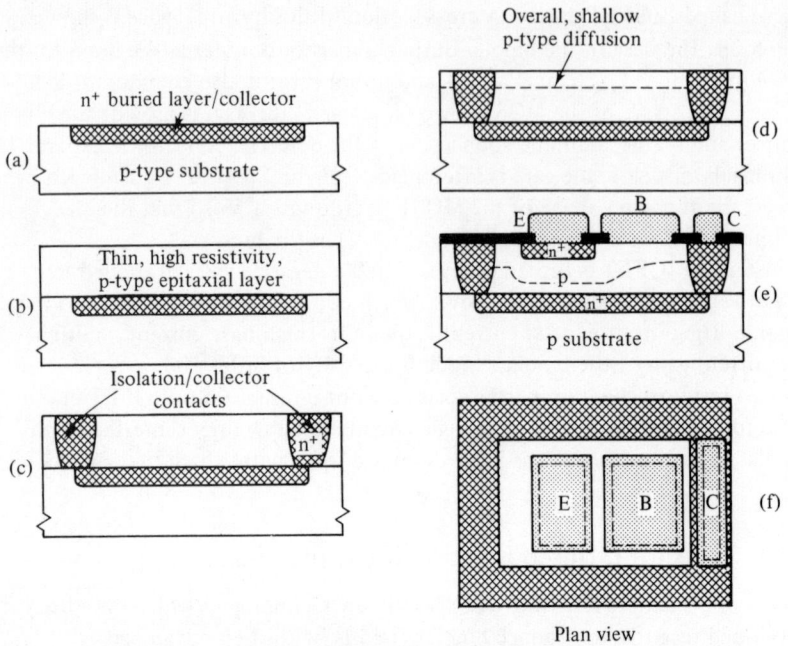

Fig. 5.29 *Stages in the manufacture of an integrated CDI transistor.*

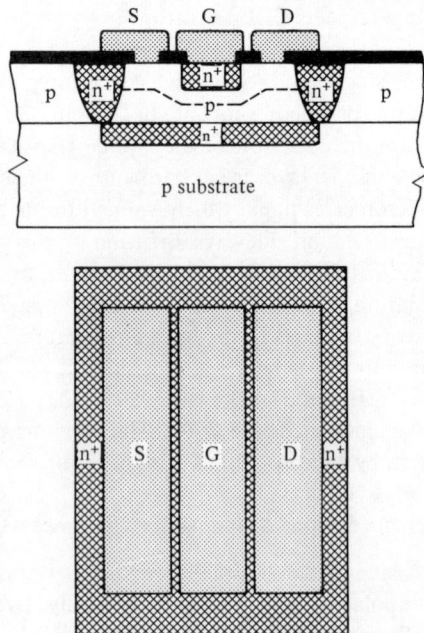

Fig. 5.30 *Collector diffused isolation JFET.*

86

(b) A device area which is typically only about one-third that of a conventional component, because of the reduced area required for isolation and collector contact, leads to an increased packing density. There is also a considerable saving in the area of n^+ diffused cross-under connections (see section 6.2).

(c) CDI transistors have a faster switching speed than most conventional bipolars, because of the reduction of charge storage at forward biased junctions, which is inhibited by the high doping levels in the collectors. An additional emitter is often included and connected internally to the base. This acts as an effective collector, which removes injected electrons from the forward biased collector-base junction and so reduces the stored charge even further.

(d) CDI transistors have a high h_{fe}, around 150, at low and high currents, a low collector access resistance and saturated resistance, due to the collector contact/isolation diffusion completely surrounding the transistor. The value of V_{CEsat} is approximately halved, which becomes important in some switching circuits.

(e) Finally, it is possible for JFETs to be produced by similar technological steps, as shown in Fig. 5.30, and incorporated on the same large-scale integrated circuit chip as bipolar devices. These field effect devices operate in the bulk of the material and are not dependent on surface effects, as are, for example, certain MOSTs.

An example of a commercially available CDI integrated circuit is illustrated in Plate 5.11 on page 88.

Problems

5.1   A linearly graded $p-n$ junction capacitor has a capacitance of 10 pF at a reverse bias voltage of 1 V. Determine its capacitance when the reverse bias voltage is changed to 5 V, assuming a contact potential of 0.8 V. What is the implication of this result?

What factor limits the maximum capacitance possible from this component?

Ans. 6.8 pF.

5.2   A particular linearly graded $p-n$ junction in silicon has a cross-sectional area of 1 mm^2 and a capacitance of 300 pF when the applied reverse bias voltage is 10 V. Calculate (a) the change in capacitance for a 50 per cent reduction in bias voltage and (b) the maximum field in the depletion layer with the 10 V reverse bias applied. Assume a relative permittivity for silicon of 12.

Ans. (a) 78 pF   (b) 4.2×10^7 Vm^{-1}.

5.3   The collector-base junction of a transistor is made by diffusing boron into a background concentration of 10^{22} atoms/m^3. The surface concentration of the boron diffusion is 10^{24} atoms/m^3 and the junction depth is 3.5 μm. Determine

87

Plate 5.11   Microphotograph of part of a silicon slice, measuring 4.5 × 3.0 mm, which contains an integrated circuit industrial timer, produced by the collector-diffusion isolation process. Reproduced by kind permission of Ferranti Ltd, Electronic Component Division.

the junction capacitance per unit area for a reverse bias of 10 V, assuming a linearly graded structure. Assume that the relative permittivity of silicon is 12.

Ans. 7.5 pF $(mm)^{-2}$.

5.4   A certain *n–p–n* transistor has the following parameters under certain bias conditions:

effective device cross-sectional area = 10^{-9} m^2
emitter current   = 100 μA
emitter-base depletion capacitance = 5 pF
base-collector depletion capacitance = 1 pF
collector access resistance   = 100 Ω
effective length of base region   = 2 μm
saturation velocity of electrons   = 8.5×10^4 ms^{-1}
relative permittivity   = 12
diffusion coefficient for electrons = 0.003 m^2 s^{-1}.

Estimate the total delay time for the device and hence its alpha cut-off frequency, at room temperature.

Ans. 2 ns; 500 MHz.

5.5   A monolithic transistor has the following specifications:

collector resistivity	= 2.5×10^{-3} Ω-m
emitter depth	= 1.5 μm
base depth	= 2 μm
substrate depth	= 15 μm
emitter area	= 50 $\times$ 8 μm
collector contact area	= 75 $\times$ 8 μm
number of collectors	= 2
distance between emitter and collector edge	= 12 μm.

Estimate the collector parasitic resistance, assuming a buried layer having a sheet resistance of (a) 15 Ω/square and (b) 5 Ω/square.
   Repeat the problem with an epitaxial layer thickness of 5 μm.
   Draw any conclusions from the results.

Ans. (a) 40 Ω (b) 39 Ω; 14 Ω; 13 Ω.

5.6   Show that the resistance between planes AB and CD of the resistive film illustrated in Fig. 5.7 is given by eq. (5.7) where ρ_s is the sheet resistance of the resistor.

5.7   A ladder network as shown in Fig. 5.31 is to be produced in integrated form, using diffused resistors and capacitors.
Sketch a possible layout for the circuit and calculate the dimensions of the com-

ponents, assuming a resistor path width of 10 μm, a sheet resistance of 500 Ω/square and a capacitance per unit area of 0.1 pF $(\mu m)^{-2}$.

Ans. 100 μm; 200 μm; 1000 $(\mu m)^2$.

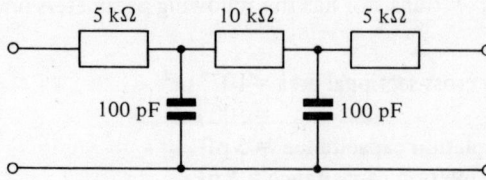

Fig. 5.31   Circuit for problem 5.7.

5.8   Sketch a possible set of masks that could be used to produce the circuit of the previous problem.

6. Electrical isolation, interconnections, and packaging

In this chapter, the various methods available for obtaining the necessary electrical isolation between circuit components, interconnection systems to provide conducting paths between components, and the final wiring and packaging operations to produce a marketable integrated circuit, will be discussed briefly.

6.1 Electrical isolation

In a fully integrated circuit, it is necessary for there to be complete electrical isolation between constituent components, the only conducting path between them being made by deliberately introduced metallic interconnections.

One obvious method of providing isolation in planar monolithic circuits is to surround individual components by a thick dielectric insulating layer, usually silicon dioxide; this so-called *dielectric isolation* is illustrated in Fig. 6.1. The

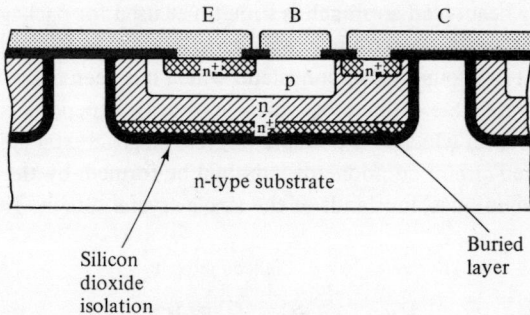

Fig. 6.1 Dielectric isolation of integrated circuit components.

advantages of the method are that there is only a relatively small stray capacitance between components and substrate, the leakage currents between components is small and no isolation biasing voltage is required. An obvious disadvantage of the method is the added number of processing steps and complexity, with a consequent reduced yield and increased cost.

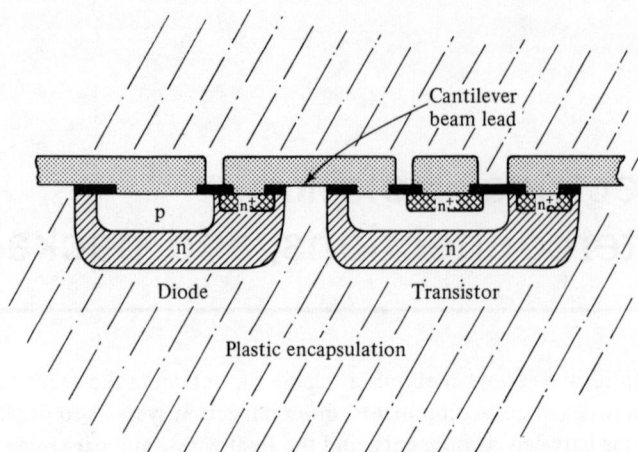

Fig. 6.2   Beam-lead integrated circuit.

Another possible method of isolation is to physically separate circuit components, using the *beam-lead* system illustrated in Fig. 6.2, to provide the highest degree of electrical isolation. The integrated circuit is manufactured in the usual way, but the aluminium interconnections are built up to relatively massive proportions, typically to around 10 μm thick, by additional vacuum evaporation. The bulk p-type substrate is then removed by etching from the back of the slice, leaving cantilever beam leads to support as well as interconnect the structure. The circuit obtains its final mechanical rigidity by being potted in thermosetting plastic. Again the extra processing steps and cost have prevented the wide adoption of this isolation technique for all but a few specialized applications, although, as will be seen, the beam-lead approach is sometimes used for packaging connections.

It is possible to provide extremely efficient electrical isolation by forming each component or group in a silicon island which has been deposited on an insulating substrate. One of the most suitable, though expensive, substrates is artificial sapphire, on which it is possible to grow single-crystal epitaxial layers of silicon. Integrated circuit components can then be formed, by the usual masking and diffusion techniques, in islands of the single-crystal silicon. This process lends

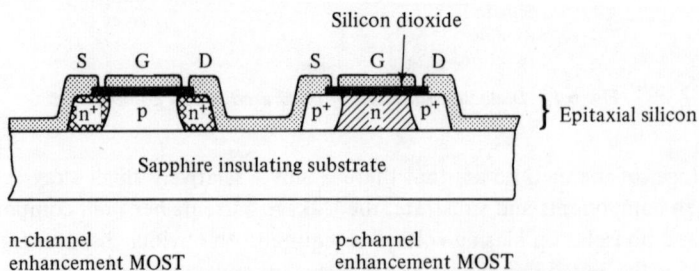

Fig. 6.3   Silicon-on-sapphire complementary MOSTs.

itself particularly well to the production of complementary pairs of devices on the same substrate, as illustrated by the complementary MOS transistor structure shown in Fig. 6.3. Cost is again a limiting factor preventing universal adoption of the silicon-on-sapphire technology, and its use is limited to special applications, such as microwave circuits.

The most common method of component isolation, *junction isolation*, is perhaps electrically the most inconvenient so far discussed, but its advantages of technological simplicity, reliability, and economy, far outweigh any disadvantages. Elements of the method, which, incidentally has been tacitly assumed for most of the integrated circuits and components discussed previously, are shown in Fig. 6.4. Each circuit element, or sometimes group of elements, is formed in an island of *n*-type epitaxial silicon, surrounded by a sea of acceptor diffused and substrate *p*-type material. Each component is then only effectively connected to its neigh-

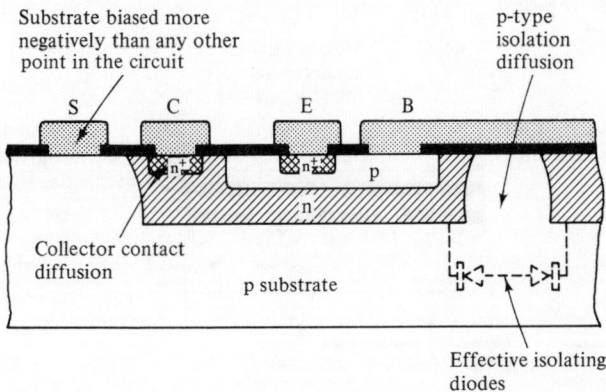

Fig. 6.4   Junction isolation in a planar monolithic integrated circuit.

bour by two back-to-back diodes, as shown, and if the *p*-type substrate is maintained at a more negative potential than the collector regions, then the electrical separation is achieved by the high resistance paths due to the reverse biased diodes. An obvious electrical disadvantage of the system is that additional parasitic capacitances which interconnect components are introduced, due to the reverse biased *p–n* junction. There also exists a parasitic *p–n–p* transistor, formed between base, collector, and substrate layers.

6.2 Interconnections, cross-overs, and cross-unders

Metallic interconnections between component parts of monolithic integrated circuits are usually formed, as discussed, by vacuum deposition of aluminium films, which make contact with the diffused circuit via windows etched in a covering oxide layer. The aluminium deposition is usually by evaporation from a heated tungsten filament loaded with pure aluminium wire in an evacuated vessel, usually a bell jar, which is maintained typically at a pressure of 10^{-4} P ($\approx 10^{-6}$ torr), to produce films of say 0.5 mm thickness. The aluminium is *micro-alloyed* to the silicon exposed through the windows by heating at around 600°C, the melting point of the silicon and aluminium eutectic alloy. Aluminium makes good ohmic contact to *p*- or n^+-type silicon but contacts to lightly doped *n*-type regions, such as collectors, usually exhibit diode characteristics. It is therefore necessary to diffuse an additional n^+ diffusion intermediate between the aluminium contact and the collector, to ensure a gradual ohmic connection, as shown, for example, in Fig. 6.4.

The final interconnection pattern is achieved by etching the surplus aluminium away via windows in a photo-resist emulsion which is spun on to the aluminium layer. The positive photoresist material which, as described in section 3.3, is used for this operation, also ensures a finer edge definition to the connections, which are typically 10 μm wide for signal wires but wider for supply lines. Bonding pads,

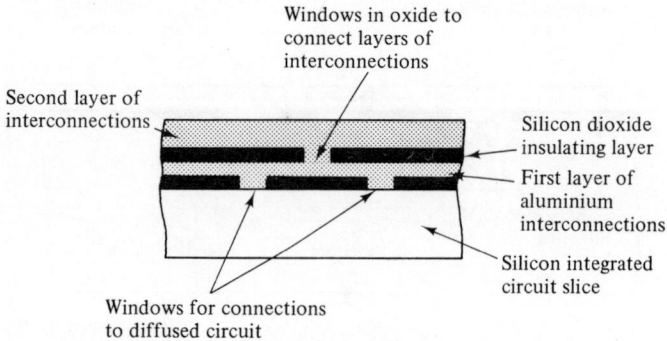

Fig. 6.5   *Multiple layer metallization.*

which terminate the interconnections and to which wires are eventually bonded to connect the circuit to external leads on the package, are also included at this stage. The aluminium pads are typically 100 x 100 μm.

It is often topologically difficult and sometimes impossible to devise a pattern of interconnecting conductors which does not include unwanted electrical connections, since, for example, connecting leads cannot be superimposed directly on top of or across each other. Computer-aided-design programs are often incorporated in the layout and mask design stages to assist in the avoidance of such spurious cross-over connections. In many complex circuits, however, conductor cross-overs are unavoidable, so various techniques have been devised to ensure that they are electrically non-contacting.

One obvious, though technically difficult, solution is to use a multiple layer metallization process, each conducting layer being physically separated by an insulating one, but being electrically connected via windows in the insulation, as shown schematically in Fig. 6.5. Such systems, using a silicon dioxide insulating layer between two or more aluminium interconnection levels, are used for some

94

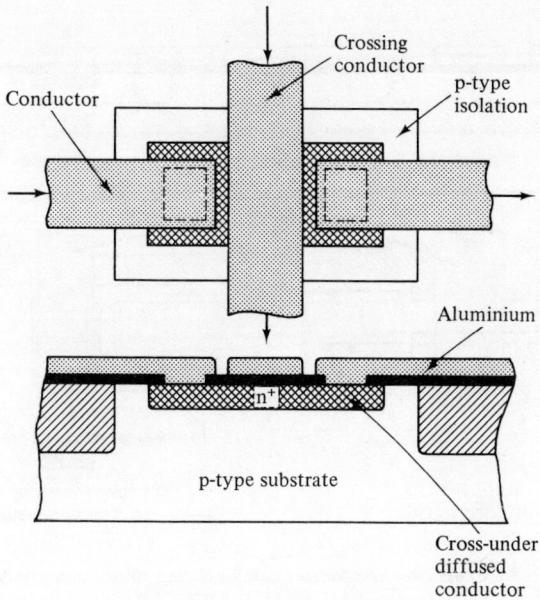

Fig. 6.6   *Diffused cross-under.*

large scale integrated MOS circuits. These are designed and manufactured using computer control and can include some degree of customer sub-system design at the interconnection mask stage. But such techniques suffer from the disadvantage of being expensive, both in capital outlay on computers, and also because of possible reduced yields due to the increased complexity in processing.

Insulated conductor intersections can also be formed by a *diffused cross-under* technique, illustrated in Fig. 6.6. A specially prepared, heavily doped n^+ region

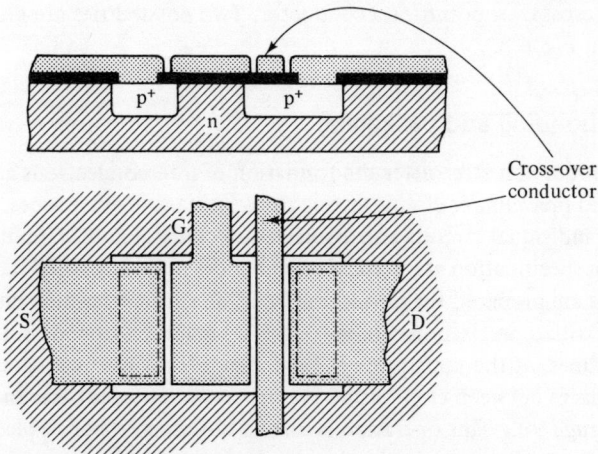

Fig. 6.7   *Extended drain diffusion used as a cross-under conductor.*

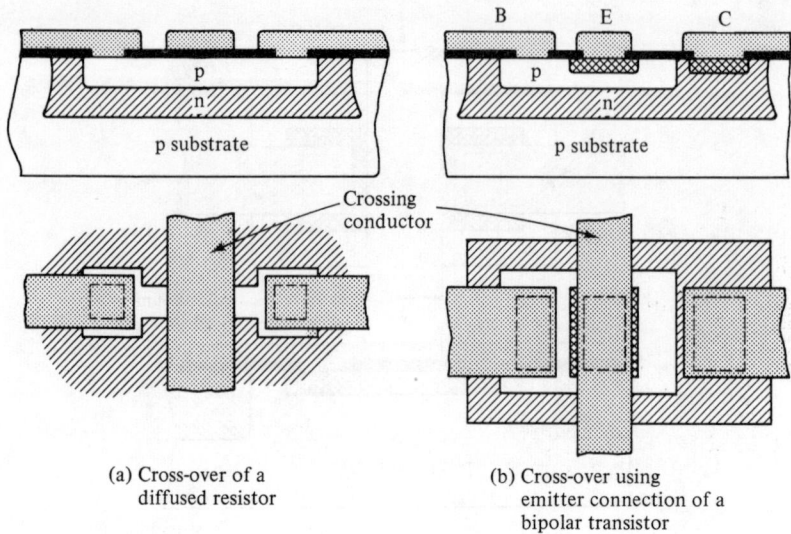

Fig. 6.8   *Cross-over conductors using existing circuit components.*

behaves as a low resistance path which forms part of one conductor, and an oxide layer effectively insulates it from the second crossing conductor. Although no extra processing steps are required, such cross-unders are demanding of chip area and are in consequence expensive, so they must be used as sparingly as possible.

It is sometimes possible to use an existing diffused layer, or a designed extension of it, as a cross-under, as shown, for example, in Fig. 6.7. Here an extended drain diffusion in a *p*-channel MOST is used as a cross-under diffusion.

Alternatively, the layout of a particular existing component can be arranged so that in conjunction with its protective insulating oxide layer, it can form a convenient cross-over point for a conductor. Two possibilities are illustrated, as examples, in Fig. 6.8.

6.3 Wire-bonding and packaging

The next production stage after the formation of interconnections and bonding pads involves preliminary circuit testing using a set of needle probes, which contact the pads of individual circuits and associated circuitry. Faulty circuits or those not meeting specification are masked on the slice, to be discarded later. Individual circuit chips are prepared for separation from the slice by scribing lines between the circuits with a precisely machine-controlled diamond-tipped tool. A series of equispaced lines, of the same pitch as the individual circuit spacing, is scratched along the spaces between circuits and the process is repeated after the slice is rotated through 90°. This operation is carried out on a rubber or plastic diaphragm, which, when stretched, breaks the slice along the scribed lines, to separate the individual circuits, in a manner analogous to the cutting of window glass. The circuit chips are delicate, small and susceptible to abrasion, so they are often handled at this and subsequent stages using a vacuum chuck.

The final series of production steps involves bonding the circuit chips to a header in a suitable package, connecting the circuit bonding pads to the external leads of the package, and encapsulating the whole unit. These processes are relatively very expensive, often comprising around 90 per cent of the total production costs, because of the highly skilled labour involved to carry out fairly slow tasks

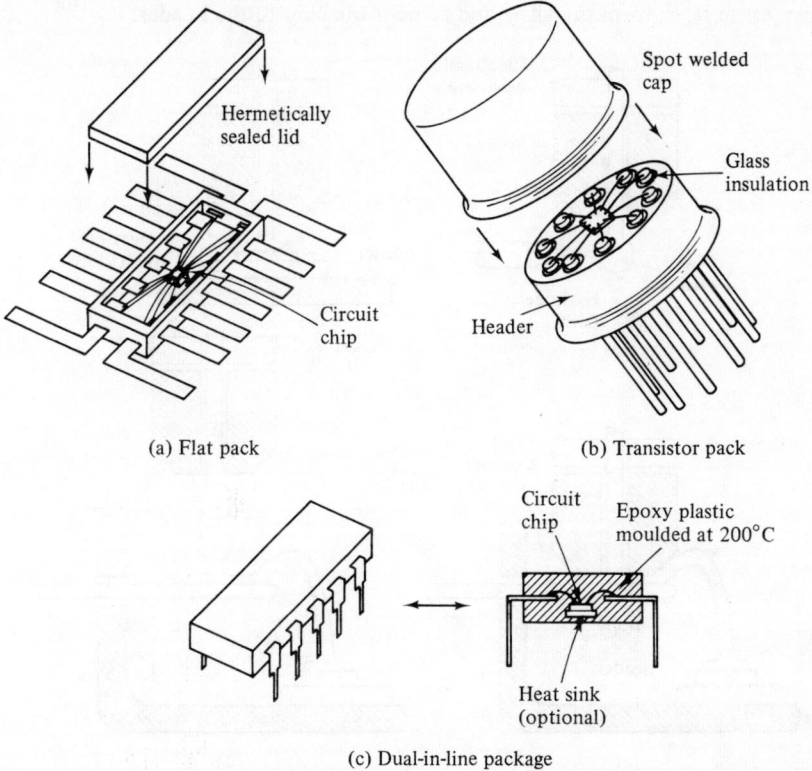

(a) Flat pack

(b) Transistor pack

(c) Dual-in-line package

Fig. 6.9   Integrated circuit packages.

and because of package costs, which can be more expensive than the circuit chips they contain!

The choice of a suitable package is wide and is dependent on the application and quality of the finished product. Some of the various possibilities are shown in Fig. 6.9. The flat pack, which was developed for aerospace applications, occupies the smallest space and is usually hermetically sealed. One difficulty is that, once fixed in position, it is difficult to remove from circuits for servicing. The transistor TO5 header, shown in Fig. 6.9(b), is a more economical alternative. Again this is difficult to solder into circuit boards by conventional flow-solder techniques because of the close spacing of its 8 or 10 leads. These considerations have led to the development of the dual-in-line package, shown in Fig. 6.9(c), which is ideally suited for printed circuit board applications. It also has the advantage of being suitable for plastic encapsulation, as shown, with the attendant advantages of ruggedness, reliability, and economy.

The circuit chips are securely fixed to the header of whatever package is chosen, usually using a gold/silicon alloy. The gold plated header is heated on an anvil to around $400°C$ and the silicon circuit chip, held in a vacuum chuck, is scrubbed down on to it, to form the alloy and cement the chip to the header.

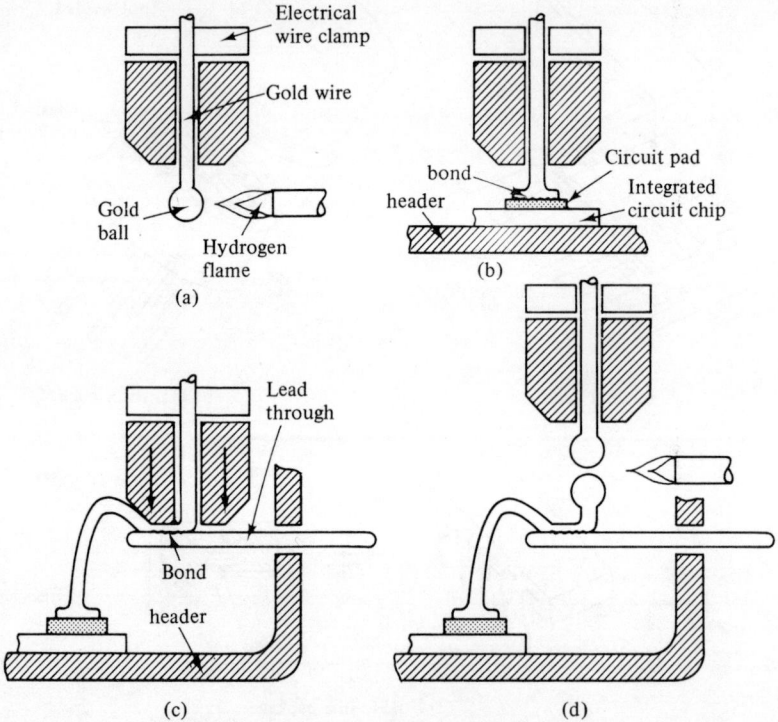

Fig. 6.10   Thermo-compression wire bonding.

The contact pads on the circuit are then connected to the external leads on the package by a wire bonding technique. The basic steps used for gold wire *ball-bonding* are shown in Fig. 6.10. A 20 μm diameter gold wire, which is fed down a capillary tube, has a 100 μm diameter gold ball formed on its end by heating with a hydrogen flame. The ball is pressed on to a circuit pad, which is preheated along with the entire circuit and header to $300°C$, to form a *thermo-compression bond*, which is a solid weld under pressure, even though the melting point of the joining metals is not approached. The wire is then released and the capillary tube raised and moved to the header post, where the wire is once again pressure-welded to the heated post, Fig. 6.10(c). The tube is finally raised and the wire is cut by a

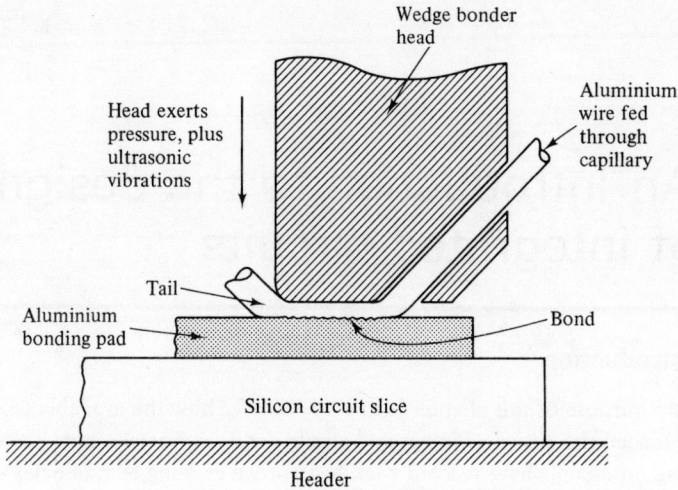

Fig. 6.11   Ultrasonic wedge bonding.

hydrogen flame, which completes one wire bond, and simultaneously forms the
gold ball required for the next one.

In an alternative form of wire bonding, ultrasonic *wedge bonding*, the bonding
wire, usually aluminium, is fed down a capillary drilled in a wedge-shaped bonding
head, as shown in Fig. 6.11. The weighted wedge squashes the bond wire down
on to the circuit bonding pad and the weld is made at room temperature but
assisted by ultrasonic vibration of the tool. Wire bonding to external leads then
progresses in the same way as for ball-bonding.

Another possibility is to use massive beam leads, as shown in Fig. 6.12, to
connect to a package or sometimes to an external circuit directly.

After all the wire bonding connections have been completed, the package,
depending on its type, is either hermetically sealed with a dry nitrogen environment, or is completely encapsulated in a thermo-setting epoxy plastic. The end
product is then subjected to a series of rigorous final electrical and mechanical
evaluation tests before being released for sale.

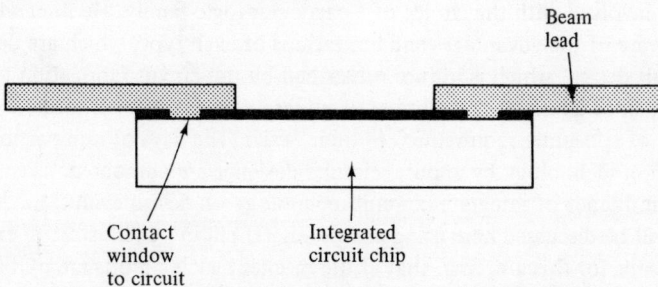

Fig. 6.12   Monolithic integrated circuit with beam leads.

99

7. An introduction to the design of integrated circuits

7.1 Introduction

The main purpose of this chapter is to study briefly how the available technology has influenced the design of integrated circuits and, conversely, how circuit and system requirements have reacted back to improve existing technologies and create new ones.

The designer of conventional discrete circuits has always been able to specify components, both active and passive, from a wide range, and covering a large spread of values and tolerances. Modern designers have less flexibility in their approach to integrated circuit design. They are more inhibited by the restricted types of component that are suitable for integration and by a possible wide spread in their tolerances, as discussed in chapter 5. Integrated circuits have to be designed to accommodate these and other constraints. For instance, there is usually an over-riding economic requirement to minimize the chip area per circuit function; this criterion, for example, may be influential in a decision to replace an otherwise acceptable circuit employing diffused load resistors by a more complex circuit which uses only active devices, but which occupies a much reduced area of silicon.

Another economic consideration which impinges on the design field is that of yield. If an integrated circuit can be designed which eliminates one masking stage, say, then the increased yield will often offset any additional circuit complexity, and a slightly inferior performance might even be acceptable sometimes.

The systems engineer may feel that he need never be involved with either the detailed design or technology of integrated circuits, but he will certainly at some time be involved with the choice of a particular logic family. He then will need to be aware of the advantages and limitations of each type, which are dependent on circuit design, which is in turn prescribed by the circuit fabrication technology. Further, since an increasing number of integrated circuits incorporate a certain amount of consumer requirement in their design, the days of non-participation in fabrication technology by a 'pure' circuits designer are numbered.

The influence of integrated circuit technology on design evolution, and vice versa, will be discussed here using mainly digital circuits as illustrative examples. The reasons for this are, first, that at the moment such circuits are by far the most commonly used and, second, because the function of a particular digital circuit is usually relatively easily and briefly explainable.

The design of a high-gain d.c. amplifier will then be discussed, as a typical example of the development of a modern linear integrated circuit which is compatible with the planar technique.

Finally, the basic operating principles and construction of a charge-coupled device will be explained, as an example of a totally integrated electronic system which could not be realized by any means other than monolithic integrated technology.

7.2 Digital integrated circuits

Currently, the most common monolithic integrated circuits are those which perform digital operations. The reasons for this trend are twofold. First, digital circuits can be designed which are extremely tolerant of quite severe variation in component values, since they are only required to process discrete signals of 1 or 0. As has been discussed in chapter 5, such component tolerances, for example a minimum tolerance of ± 20 per cent for planar diffused resistors, are an inherent feature of the planar integrated process, so it is important that these variations can be accommodated in a particular circuit design. Second, digital systems require large numbers of a few types of basic circuit, which because of the simultaneous processing of many similar circuits on a slice, using the photolithographic processes described in chapters 1 and 3, makes them particularly suitable for the planar monolithic fabrication process.

The general choice of a particular circuit is based in part on optimization of the following design criteria:

(a) the number of logic gates per unit area of chip, or the *function density*, should be maximized.
(b) the speed x power product for a particular operation should be as large as possible.
(c) the circuits should possess as high an immunity to noise as can be achieved.

The evolution of the more basic integrated logic circuits and the influence of particular fabrication technologies on circuit design will be discussed in the following sections.

7.3 Direct coupled transistor logic (DCTL)

In this, one of the earliest and simplest families of integrated logic circuits, coupling capacitors are eliminated by direct connections between transistor stages. Avoidance of the use of such capacitors is an advantage in integrated circuits, because of the difficulties in providing any substantial capacitance using a diffused capacitor of economic size, which is limited to about $100 \, \text{pF mm}^{-2}$. Similar difficulties of excessive occupation of chip area, expense and yield are experienced if an MOS or thin-film component is substituted, as discussed in section 5.1.

Examples of some of the basic DCTL logic circuits are shown in Fig. 7.1. A DCTL inverter circuit, which is the primary member of the family from which all the other circuits are derived, is shown in Fig. 7.1(a). When a positive voltage pulse, designated a logic 1, is applied to the base of the input transistor, T_1, the transistor is turned on and its collector voltage, by virtue of the voltage drop in the load resistor, R_L, falls from around the supply line voltage, V_{CC}, to the saturation voltage, V_{CEsat}, which is typically 0.1–0.2 V. This voltage is directly connected to T_2, which switches off, producing a positive pulse, which is applied to the

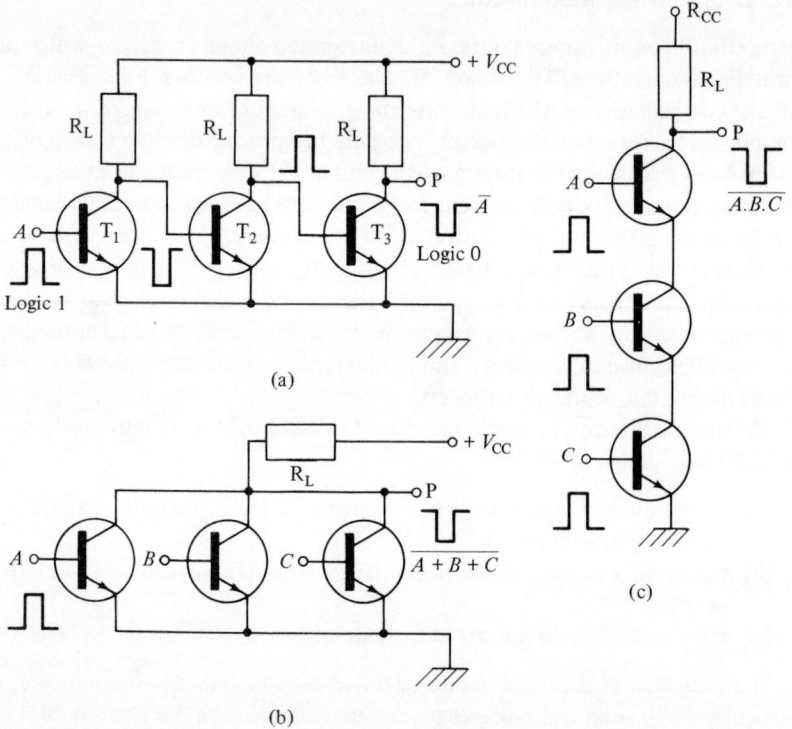

Fig. 7.1   *Principal members of the DCTL logic family. (a) an inverter, (b) a three-input NOR gate, and (c) a three-input NAND gate.*

base of T_3. This output transistor is switched on and a negative pulse, designated a logic 0, appears at the output terminal. The input positive pulse, a logic 1, has been *inverted* to a negative output pulse, a logic 0. One obvious requirement for the transistors in such a circuit is that they should possess a low value for the saturated voltage when they are turned on hard, V_{CEsat}, so ensuring that subsequent directly coupled stages are fully turned off. This condition can be achieved using bipolar integrated monolithic transistors, described in section 5.4.

A DCTL NOR logic gate is shown in Fig. 7.1(b). This consists essentially of several inverters, in this example three, in parallel, with a common load resistor.

If a positive voltage pulse ($\equiv 1$) is applied to the input to T_1, A, *or* to B or C, base current is drawn and one or more of the *n–p–n* transistors is turned on. The common collector voltage, V_0, falls from V_{CC}, ($\equiv 1$), to V_{CEsat}, ($\equiv 0$). Hence the logic function of the circuit is described by:

$$A + B + C = \overline{P} \tag{7.1}$$

i.e., a logic 1 at input A or B or C produces an inverted logic 1, i.e., a logic 0, at the output P.

Figure 7.1(c) shows a three-input DCTL NAND gate, formed from inverters in series. Only when all three transistors are switched on, by the application of a logic 1 pulse to A *and* B *and* C, does current flow through R_L and the output at P falls to $3V_{CEsat}$, equivalent to a zero. This logic function is described by:

$$A . B . C = \overline{P} \tag{7.2}$$

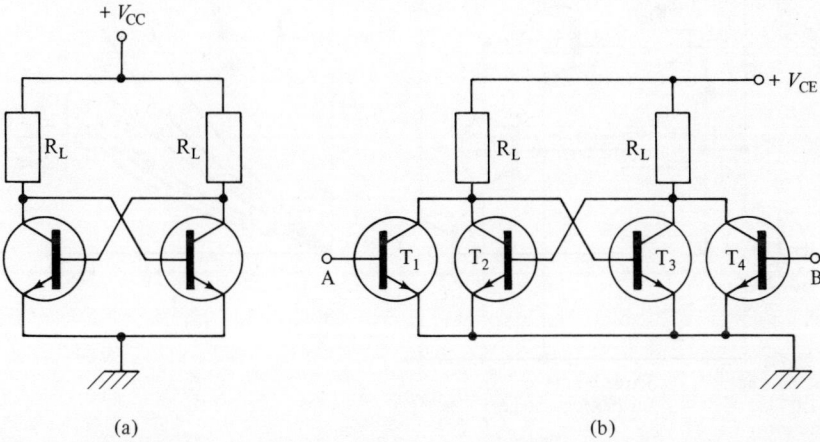

Fig. 7.2   DCTL bistable circuits (a) the basic circuit and (b) the practical circuit including switching transistors.

A DCTL bistable or flipflop circuit can be made by cross-connecting two inverters as shown in Fig. 7.2(a). In a practical circuit, additional triggering transistors are included, as shown in Fig. 7.2(b). If a positive pulse is applied to one of the triggering transistors, T_1 say, then T_2 is switched off and its bistable partner T_3 is switched hard on. The states of T_2 and T_3 can then be interchanged by the application of a pulse to the other switching transistor T_4, at B.

DCTL integrated logic circuits have the distinct technological and economical advantages of being simple, requiring few components, only needing one power supply and, because of the absence of coupling capacitors, chip area is conserved. Further reductions in area are possible in particular circuit configurations; for example, all collectors are common in the DCTL NAND circuit, Fig. 7.1(b), so all the transistors can be formed in one isolation island, with a considerable improvement in packing density.

However, DCTL circuits suffer from the disadvantages of a limited switching speed, due to the storage of charge when transistors are driven hard on into saturation, and their susceptibility to noise cannot always be ignored. But perhaps the principal reason for their now being superseded is their requirement for relatively close-tolerance components, particularly transistors, for successful and reliable operation. This difficulty arises from the typical use of the output of one logic gate, T_1, to supply the inputs to several others T_2, T_3, T_4, illustrated in Fig. 7.3(a), a process known as *fan-out*. If transistors T_2, T_3, T_4 are not identical, then

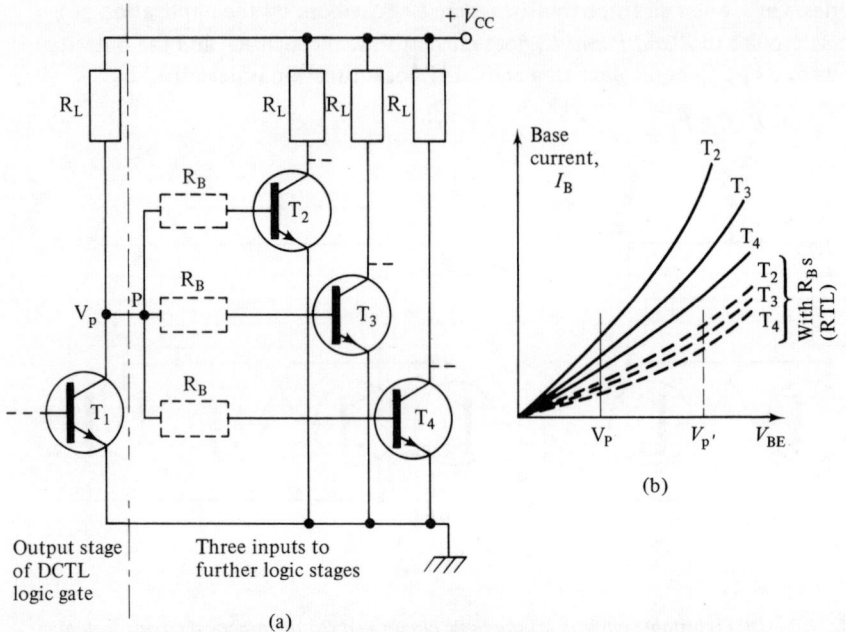

(a)

(b)

Fig. 7.3 (a) Fan-out of DCTL circuits (the R_Bs are included for RTL circuits only) and (b) possible mutual characteristics of input transistors T_{2-4}, with and without base resistors.

their base currents, supplied by the load resistor to T_1, are shared unequally, as shown in Fig. 7.3(b), which is known as *current hogging*. In this example, T_2 might then be turned ON but another, T_3, might not be, causing a failure in the logic operation.

7.4 Resistor transistor logic circuits (RTL)

These circuits were evolved to reduce current hogging effects in integrated DCTL circuits by the inclusion of diffused resistors, R_B, in series with all transistor bases, as shown dashed in Fig. 7.3(a). The input voltage to turn on transistors is then increased, to $V_{p'}$ in Fig. 7.3(b), but in other respects RTL circuits perform logic functions in an analogous manner to their DCTL equivalent. However, the base resistors do tend to equalize the input characteristics when fan-out is necessary, as shown dashed in Fig. 7.3(b); thus current hogging is virtually eliminated and noise immunity can be improved because of the wider signal swings that are possible.

Integrated RTL circuits are slower than their DCTL counterparts because of the increased switching time constant introduced by the R_Bs. The circuits can be speeded up by the use of high-speed transistors and by reducing base resistances, but this is at the expense of increased power dissipation. A further disadvantage of integrated RTL circuits is that they are relatively wasteful in chip area. For example, a 1 kΩ diffused resistor uses the same area that could be used to accommodate up to 10 bipolar integrated transistors, so, obviously, the number of resistors used per gate should be kept to a minimum. Because of these difficulties, with the advent of TTL, described later, which provides more easily realizable and economically viable integrated circuit gates, RTL circuits have now become largely obsolete.

7.5 Diode transistor logic circuits (DTL)

A typical integrated DTL logic gate is shown in Fig. 7.4. This improved version of the basic circuit is sometimes known as *low level* DTL (or LLL).

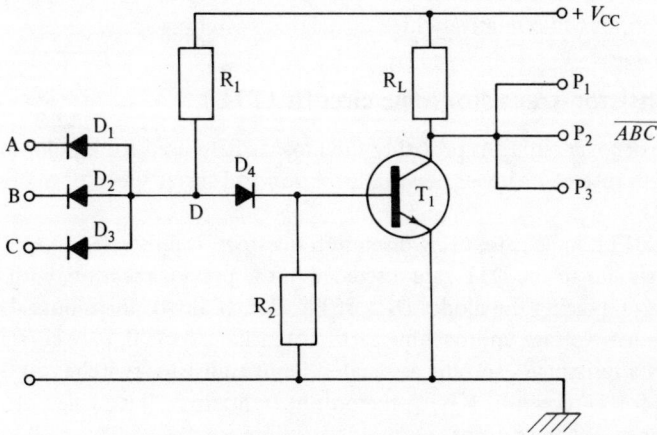

Fig. 7.4   DTL NAND gate.

Diodes D_{1-3}, in conjunction with R_1, perform the circuit's logic function, behaving as an AND gate, but the output transistor, T_1, acts as an amplifier and an inverter, and the gate overall performs a NAND operation. If any of the inputs A–C is at the saturation voltage of a previous driving stage, which is typically about 0.2 V above earth, a logic 0, then its associated diode conducts. The voltage at circuit point D then falls, because of the current drawn through R_1, to a value which is less than the base-emitter voltage of transistor T_1 plus the diode drop in D_4

and is therefore insufficient for base current to flow. Transistor T_1 is turned off and a logic 1, approximately equal to V_{CC}, appears at the outputs P_{1-3}.

If now all the inputs are simultaneously supplied with a logic 1 pulse, diodes D_{1-3} do not conduct, the voltage at circuit point D rises towards V_{CC}, base current flows and T_1 is driven hard on. A logic 0 then occurs at the output. The gate thus functions as a NAND gate described by:

$$\bar{P} = A.B.C. \tag{7.3}$$

Resistor R_L is not part of the logic circuitry but is included to provide a conducting path to earth for charge stored in the transistor to leak away when it is turned off, so decreasing the switching time.

DTL integrated logic circuits do not suffer from current hogging and fan-out to drive many succeeding gates is possible. A further advantage, which is most valuable in the integrated circuit version, occurs as a result of all the diodes, D_{1-4}, having electrically common anodes. This allows them to be located in a common isolating island, which makes possible a large number of inputs in a relatively small chip area.

The choice of diodes available using the integrated planar process is outlined in section 5.5. Usually the diode configuration shown in Fig. 5.19(d), consisting of a basic epitaxial bipolar transistor structure, with base and collector strapped together via the metallized interconnecting layer, is found to be most convenient, on the grounds of its relatively high associated switching speed.

7.6 Transistor–transistor logic circuits (TTL)

This type of logic circuit is probably the most widely used for bipolar integrated circuits, because of its lower power dissipation and faster speed of operation than DTL or LLL.

A basic TTL logic gate, using discrete transistors, is illustrated in Fig. 7.5. This circuit is similar to the DTL gate discussed in the previous section, with input transistors T_{1-3} replacing the diodes D_{1-3} of Fig. 7.4. If any of the inputs A, B, C is taken to a low voltage approaching earth potential, a logic 0, base current is supplied via resistor R_1 and the associated input transistor switches on. Its collector voltage is forced down to V_{CEsat}, the output transistor, T_4 is turned off and a logic 1 appears at the output.

Now if all the inputs are simultaneously subjected to a positive voltage pulse of sufficient magnitude, a logic 1, the input transistors behave merely as diodes through which base current flows, via R_1, to the output transistor, which turns on. A logic 0, equal to V_{CEsat}, appears at the outputs P_{1-3}.

Hence the circuit functions as a logic NAND gate described by:

$$A.B.C = \bar{P} \tag{7.4}$$

It will be noticed from Fig. 7.5 that the collectors and emitters of the input transistors are electrically common. Thus multi-emitter transistors, in which

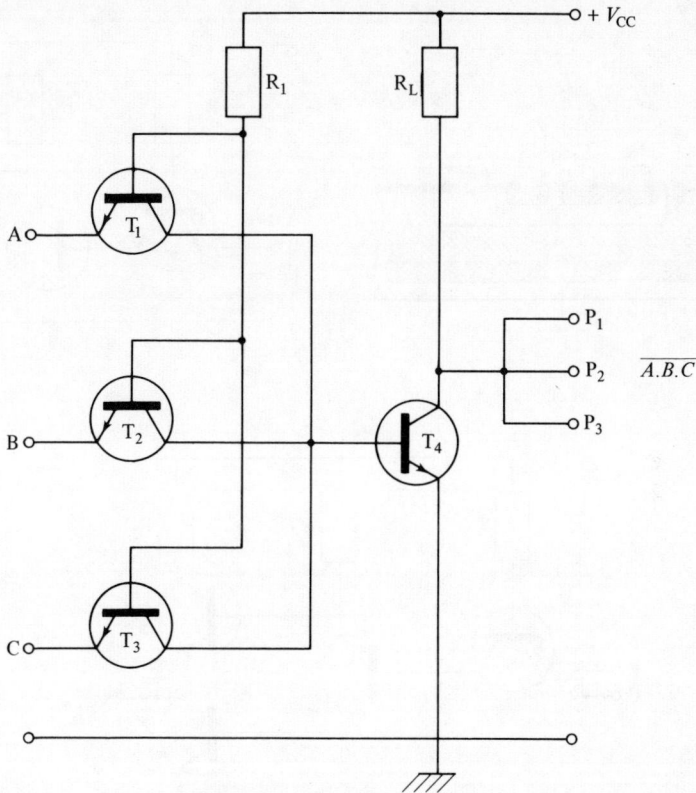

Fig. 7.5   A basic TTL NAND gate.

several emitters can be incorporated in a single base and collector structure, as shown in Fig. 7.6(c). can be incorporated to advantage in integrated TTL circuits, which leads to a valuable saving in chip area per input. Naturally, the emitters are spaced sufficiently far apart so as not to interact with each other. The three-input integrated TTL NAND gate is then reduced to the basic circuit shown in Fig. 7.6(b) which has developed into the typical integrated positive NAND gate illustrated in Fig. 7.6(a). A further advantage of TTL logic circuits is that the turn-off switching time is much reduced, since charge stored in the output transistor can leak away quickly to earth via the low impedance between the collector and emitter of an input transistor that is switched on. This results in propagation delays of around 10 ns for standard integrated TTL gates, at a dissipation per gate of, say, 20 mW. High-speed integrated TTL circuits, with delays of around 3 ns per gate, have been made, but at the expense of increased dissipation, to around 30 mW per gate. Alternatively, low-power integrated TTL is available with dissipation of 1 mW per gate, but having an increased delay time, around 30 ns.

Integrated TTL circuits also have a good fan-out capability and only require one supply voltage. However, one disadvantage of TTL logic circuits is that they

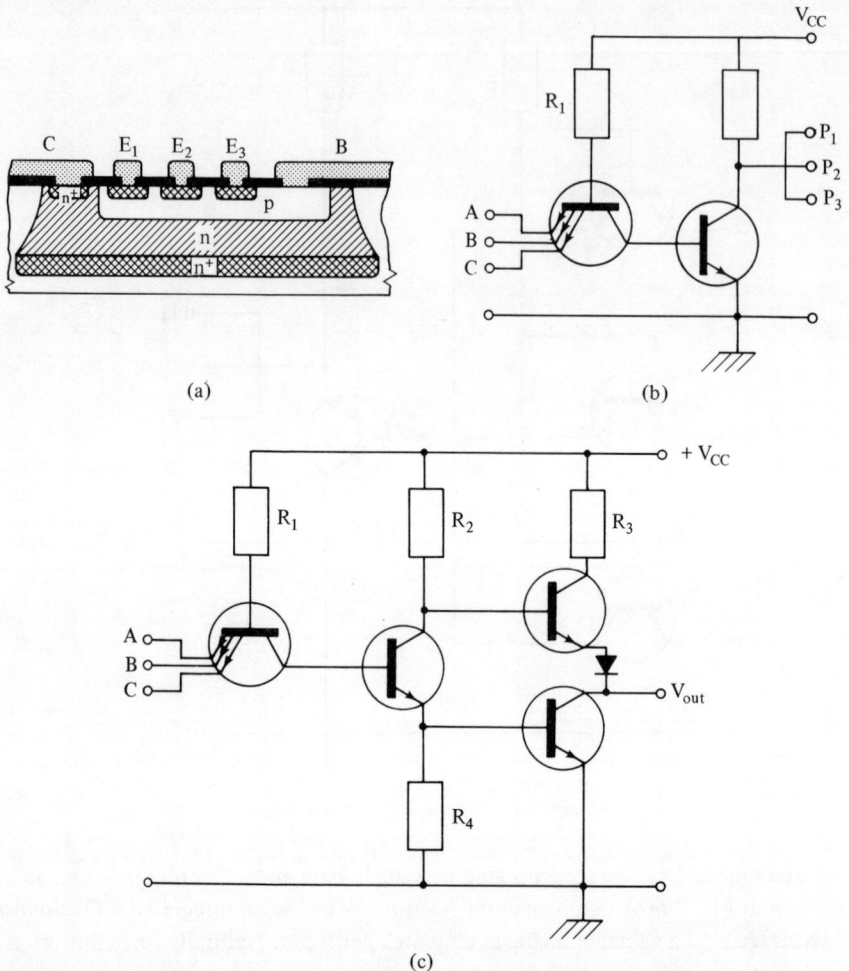

Fig. 7.6 (a) A multi-emitter integrated transistor, (b) basic integrated TTL NAND gate with multi-emitter transistor input, and (c) a typical commercial positive NAND gate.

are, in the simple version described, somewhat sensitive to noise. This arises because the output transistor only requires about 0.7 V at its base to switch on. In its off state, it is prevented from switching on by a base voltage of V_{CEsat} from an input transistor, which is often only slightly smaller than 0.7 V. Hence a relatively small additional noise voltage could cause the output transistor to switch spuriously and record a false logic signal.

A basic TTL gate, using multi-emitter CDI transistor technology, as described in section 5.11, is shown in Fig. 7.7. Although the circuit function looks basically the same as for TTL circuits fabricated using conventional planar technology (compare for example Figs. 7.6(c) and 7.7), the inherent high inverse current

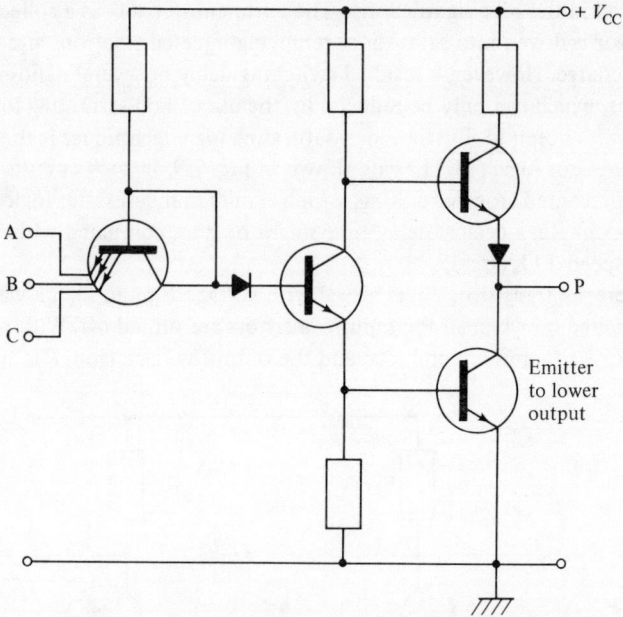

Fig. 7.7   Basic integrated TTL gate using CDI technology.

gain associated with CDI transistors prohibits the usual input arrangements used in conventional TTL circuits. The collector of the input transistor is consequently shorted to the base as shown, to produce a circuit behaviour which is more akin to DTL logic.

7.7 Non-saturating logic circuits. Emitter coupled transistor logic (ECTL)

In all the previously described logic circuits the transistors are driven hard on into saturation. In so doing they store charge, which inevitably leads to long switching turn-off times as discussed, for example, in section 5.5.1. Additional feedback emitters, which have been incorporated in some CDI integrated transistors and internally connected to transistor bases, as shown in Fig. 7.8, can reduce the

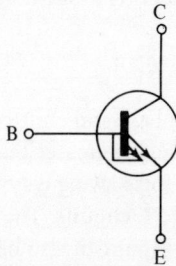

Fig. 7.8   CDI transistor with feedback emitter.

109

charge stored in the base significantly. The extra emitter acts as a collector when the transistor is driven into saturation, removing injected electrons and so reducing the stored charge. However, a residual switching delay of several nanoseconds still remains, which can only be reduced by the use of non-saturating logic circuits.

The circuit chosen to illustrate non-saturating logic techniques is the emitter coupled transistor logic (ECTL) gate shown in Fig. 7.9. In such circuits the transistors are prevented from saturating, which results in much faster logic operations. For example, a typical delay time might be 2 ns, compared with 10 ns for a medium-speed TTL circuit.

The reference transistor, T_4, is biased with voltage V_{BB}, in such a way that it is always switched on when all the input transistors are turned off. With no inputs to A, B or C, T_4 therefore conducts, and the output voltage from P is approxi-

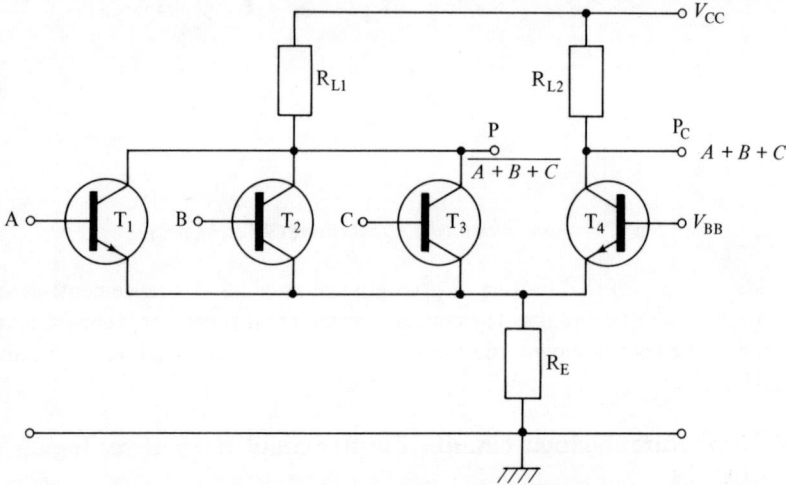

Fig. 7.9   Emitter coupled transistor logic NOR/OR gate.

mately V_{CC}, a logic 1. Now, if a positive logic input voltage pulse, $V_{in} > V_{BB}$, is applied to one of the inputs A, B or C, the corresponding transistor conducts. The common emitter resistor, R_E, which is connected to earth, or sometimes a negative supply line, behaves as a constant current source, so current is transferred from the reference transistor to the conducting input transistor. The output at P then falls to a voltage:

$$V_P \simeq V_{CC} - (V_{in} - V_{BE})R_{L1}/R_E$$

where V_{BE} is the volt drop between base and emitter of the input transistor. This voltage, defined as a logic 0, can be made greater than V_{CEsat} by a correct choice of R_{L1}, so avoiding saturation. The logic swing is typically as low as 1 V, compared to around 3 V for DTL and TTL circuits. The circuit, as described, functions as a NOR gate. A complementary output can also be obtained from the collector of the reference transistor, at P_C, and the gate then provides an OR logic function.

110

A basic integrated, non-saturating gate, which uses the CDI technology described in section 5.11 and operates in a current source logic mode is shown in Fig. 7.10. An inverted CDI transistor, T_4, which replaces the common emitter resistor, is biased to operate as a linear current source. The use of a transistor in this way, instead of a high value diffused resistor, which would otherwise be required, is an effective method of conserving chip area. Logic inputs to A, B or C switch on the respective input transistor and current is switched into the load resistor, R_L, at an extremely fast speed.

One obvious difficulty which must be tolerated with the admittedly faster non-saturating logic gates is that two supply voltages are required, which are also required to be large to keep the transistor currents and therefore the logic levels constant. A further problem with ECTL circuits is that the logic voltage level is

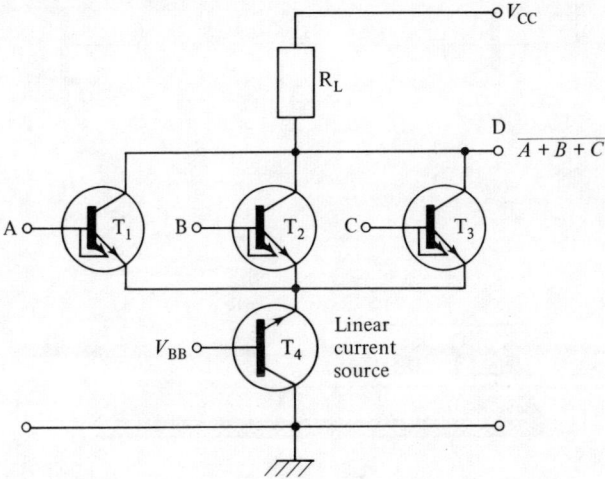

Fig. 7.10   CDI non-saturating logic gate.

not preserved, i.e., the output voltage from a gate is greater than the input voltage. This difficulty can be overcome by the inclusion of additional emitter follower circuits at the output, which provide a bonus of reduced cross-talk and an increased fan-out capability.

7.8 Schottky diode clamped TTL circuits

A variant of standard epitaxial planar technology has been developed which enables Schottky barrier diodes to be conveniently incorporated into bipolar TTL integrated circuits, to produce non-saturating, fast logic circuits. This Schottky TTL family of integrated circuits has a speed capability which is similar to that of ECL circuits, without having some of their disadvantages, such as the need for two supplies, and is compatible with and as convenient to use as standard integrated TTL.

Whereas Schottky barrier diodes have been available for some time, it is only recently, with the development of the Schottky barrier clamped integrated transistor, that they have achieved sufficient reliability to be incorporated in integrated circuits. The basic integrated component is illustrated in Fig. 7.11(a) and its symbolic representation is shown in Fig. 7.11(b). It will be seen to consist essentially of an n–p–n bipolar transistor with an integral Schottky diode connected between its collector and base, which prevents the transistor going into saturation.

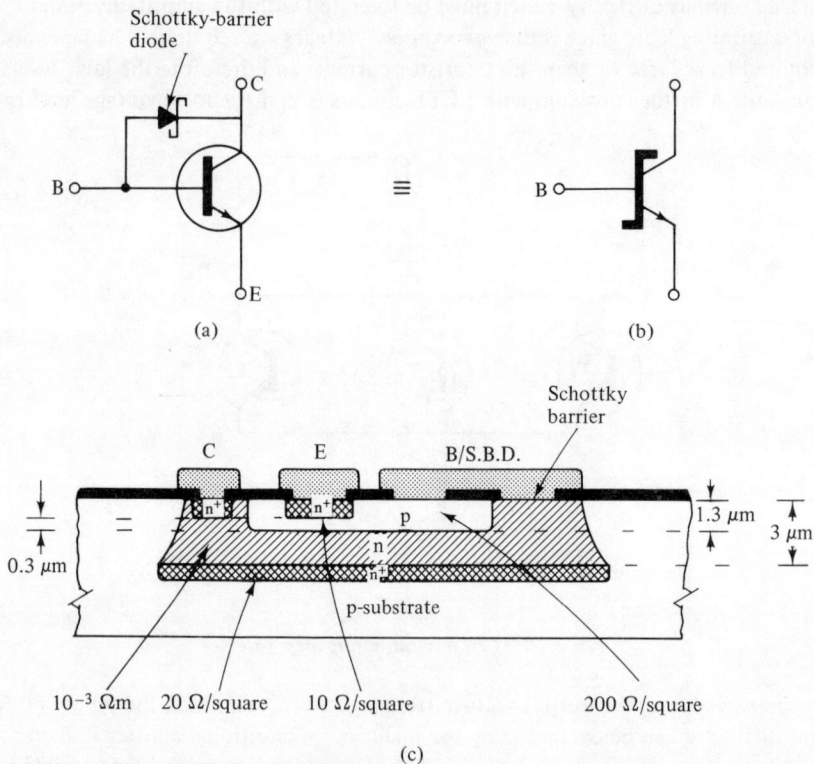

Fig. 7.11 (a) The Schottky-barrier diode clamped transistor (b) its symbol and (c) its realization in the planar integrated technology.

Saturation is prevented because of the low forward voltage drop of the diode, around 0.4 V, compared to the voltage necessary to forward bias the collector-base junction of the transistor, which is around 0.7 V. Therefore, when the transistor is driven towards saturation and its collector-base junction approaches the forward bias condition, saturation is avoided by the Schottky barrier diode which diverts current into the collector and clamps the collector-base voltage at around 0.4 V. When the base current is reversed, the collector potential rises rapidly, typically in picoseconds, because of the low storage time associated with the

112

Schottky barrier. In this way, switching delays caused by minority-carrier charge storage in saturating circuits can be considerably reduced. Gold doping, which sometimes has been introduced to increase switching speed, see section 5.5.1, is also eliminated using this technique, so improving yield.

The integrated aluminium–silicon Schottky barrier diode is produced essentially as described in section 5.5.4. However, since in its circuit configuration its cathode is common to the collector of an integrated transistor, it can be conveniently incorporated with a planar transistor, as shown diagrammatically in Fig. 7.11(c). A Schottky barrier diode is formed between the aluminium metallizing and the n-type collector diffusion, provided the usual n^+ contacting diffusion is

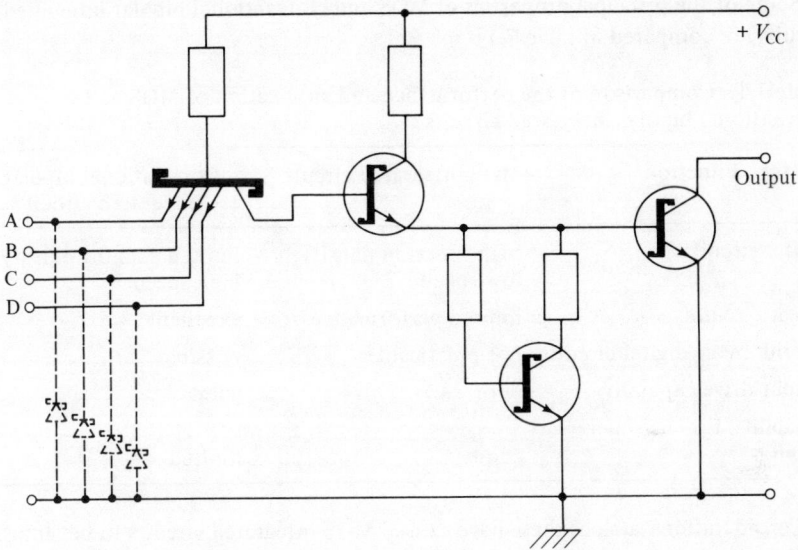

Fig. 7.12   An integrated positive NAND gate using Schottky diode clamped TTL.

omitted, as shown. The collector connection is connected to this diffusion via an n^+ emitter-type diffusion in the usual way. The diode anode and base connection are strapped internally via the metallization, as shown. Typical dimensions and resistivities have been included in the drawing, to give some idea of the latest developments of the planar process.

A typical integrated positive NAND gate employing multiple-emitter Schottky barrier clamped technology is shown in Fig. 7.12. Notice the similarity of the circuit to standard TTL. The optional input diodes limit negative voltage excursions there. As soon as the voltage at an input exceeds the forward voltage of a diode, a low impedance path is provided to the earth rail. Logic gates of this type possess propagation delays which are typically around 3 ns with a speed-power product of 60 pJ.

7.9 IGFET logic circuits

Conventional DTL, TTL, and ECL bipolar integrated logic circuits require around nine masking operations and it is becoming difficult to increase the function density for such circuits above the current limit of around 100 gates on a chip size of 1 x 1 mm. Logic circuits employing an MOS technology only require, typically, four masks, and a significant increase in gate packing density and also in possible chip size becomes feasible. Integrated MOST circuits can have higher input impedances, consume less power and occupy a smaller area than their bipolar counterpart which performs the same function. However, they are relatively slow, have a limited high-frequency performance and are not capable of driving such high currents.

Some of the principal properties of MOS and conventional bipolar integrated circuits are compared in table 7.1

Table 7.1. Comparison of the performance and application of MOS and conventional bipolar integrated circuits.

Function	MOS integrated circuit	Conventional bipolar integrated circuit
Digital circuits	high function density, low speed	limited packing density, high speed
Linear circuits	limited performance	excellent
Hybrid linear/digital circuits	not practicable	possible
Output drive capability	poor	good
Capability for customer design	good	limited

Considerations such as these have caused MOS integrated circuits to be almost exclusively adopted to medium-speed, large-scale integrated (LSI) digital circuits.

The simple bipolar DCT digital circuits described in section 7.3 have their direct equivalent in the MOS technology. Some of the basic logic gates using, as an example, induced p-channel MOSTs, are shown in Fig. 7.13. It will be noticed that the diffused load resistors of the bipolar circuit equivalents are replaced by active devices, as has been discussed in section 5.9. This is because it is much more economical in area to use a MOST biased into saturation as an effective high value load resistance, as shown in Fig. 7.13, and this has become standard practice.

The basic MOS integrated inverter, in which all the other members of the logic family have their origin, is shown in Fig. 7.13(a). The gate voltage of the load transistor, T_2, is held at the negative drain supply voltage, $-V_{DD}$. The sheet resistance of the resulting MOS resistor is about 10^4 Ω/square, which is two orders of magnitude greater than that for a diffused resistor, so a typical value of load resistance, 0.1 MΩ, can be obtained from a device with active dimensions, say, of 100 μm x 10 μm. A negative pulse, logic 1, applied to the gate of the switching

114

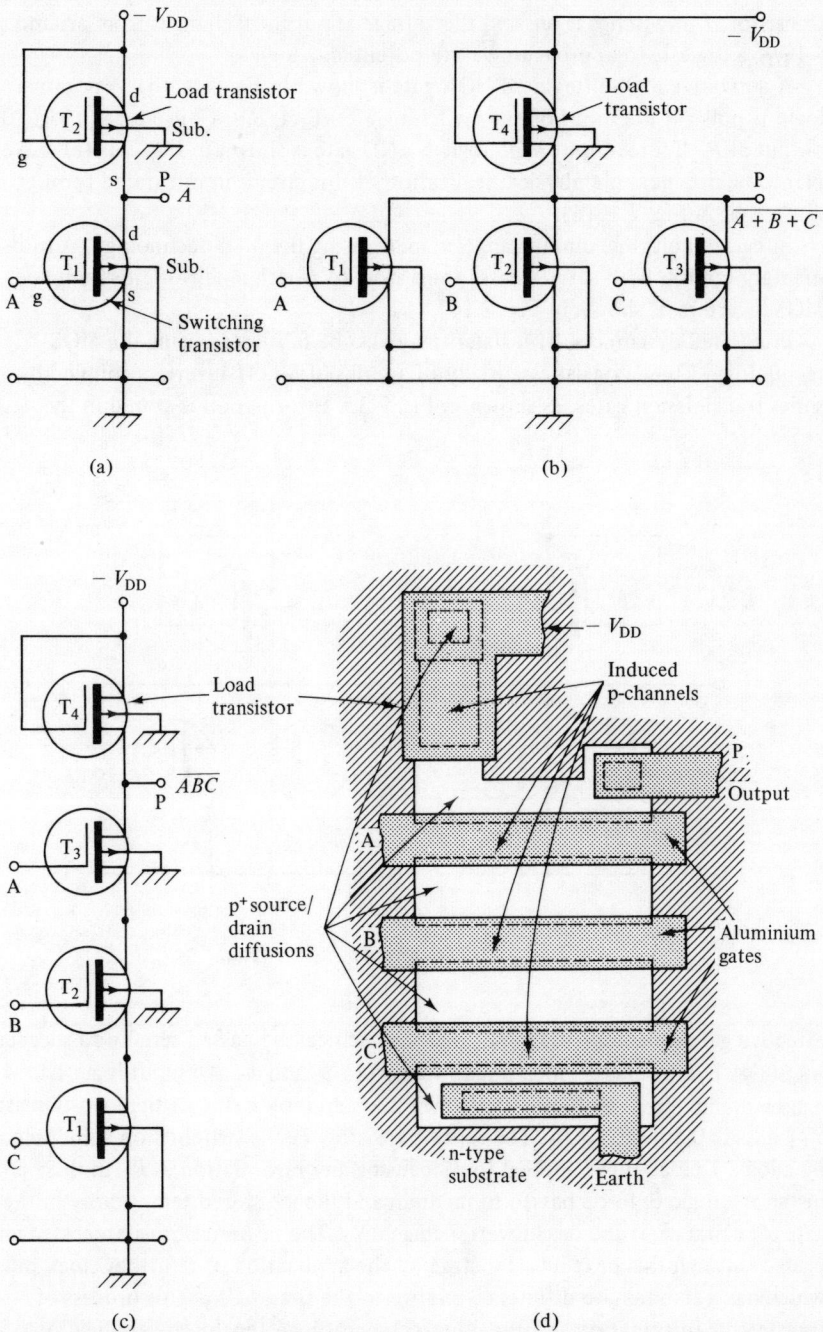

Fig. 7.13 *Integrated induced p-channel MOST logic gates. (a) an inverter, (b) a three-input NOR, (c) a three-input NAND and (d) its possible I.C. layout scheme.*

115

transistor, T_1, switches it on, and the voltage at output P changes from around $-V_{DD}$, a logic 1, to approaching earth potential, a logic 0.

A derivative MOS integrated NOR gate is shown in Fig. 7.13(b). A negative, logic 1, pulse at the input of any switching MOST, A, B or C, produces a logic 0 output at P. A three-input MOS logic NAND gate is shown in Fig. 7.13(c) and a plan view of a possible physical realization of this circuit in integrated form is illustrated in Fig. 7.13(d).

Integrated bistable circuits are also made using the MOS technology. A rudimentary circuit, with set and reset transistors to switch a pair of cross-coupled MOS inverters, is shown in Fig. 7.14.

Integrated dynamic shift registers may also be fabricated using the MOS technology. These consist essentially of a series of MOST inverters coupled by series transmission gates, as shown in Fig. 7.15. Information is stored in the

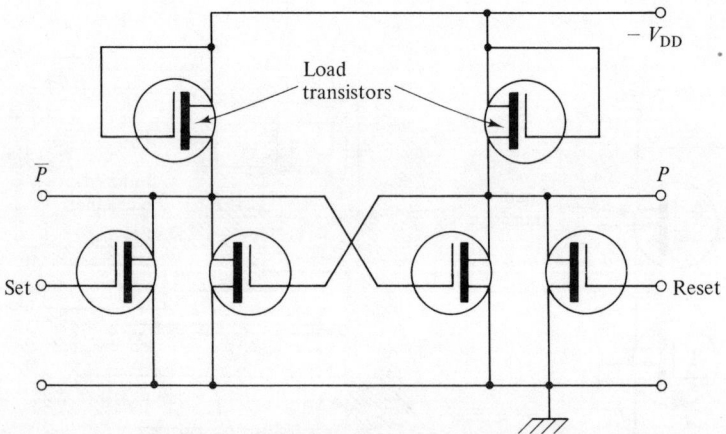

Fig. 7.14   MOS integrated bistable circuit.

effective gate capacitances C_{g2}, C_{g3} etc., which can be passed serially to succeeding stages by two-phase clock pulses applied to ϕ_1 and ϕ_2. An input logic 1 to A causes the inverter transistor, T_1, to switch on and, by virtue of the load transistor, T_{L1} a logic 0 appears at P_1. The series transistor, T_{S1} is switched on simultaneously by a logic 1 clock pulse applied to ϕ_1, causing any information at P_1, in this instance a logic 0, to be passed to its drain and thence stored temporarily in the gate capacitance of the next inverter stage, C_{g2}. The information is processed and passed through this next inverter stage by the application of a further clock pulse, which has a fixed phase difference relative to the first, to ϕ_2. This process of transferring information serially along a chain of inverters by the application of two-phase clock pulses to series coupling elements can be continued as required, to form a dynamic shift register. The maximum speed of such a register is limited

116

by the time constant associated with charging up a storage capacitor, C_g, and a minimum clock frequency, typically 10 kHz, is also determined by its discharging time constant. For clock frequencies lower than the minimum value, information is stored too long in the gate capacitance and tends to leak away.

We have seen that bipolar and MOS integrated logic systems have been developed to exploit the respective technologies. Choice of a particular logic system and technology is determined by such considerations as are summarized in table 7.1. Mixed bipolar-MOS logic circuits can also be included in large integrated systems, using, for example, the technology outlined in section 5.10. Note, however, that MOS and bipolar circuits are not always directly compatible with regard

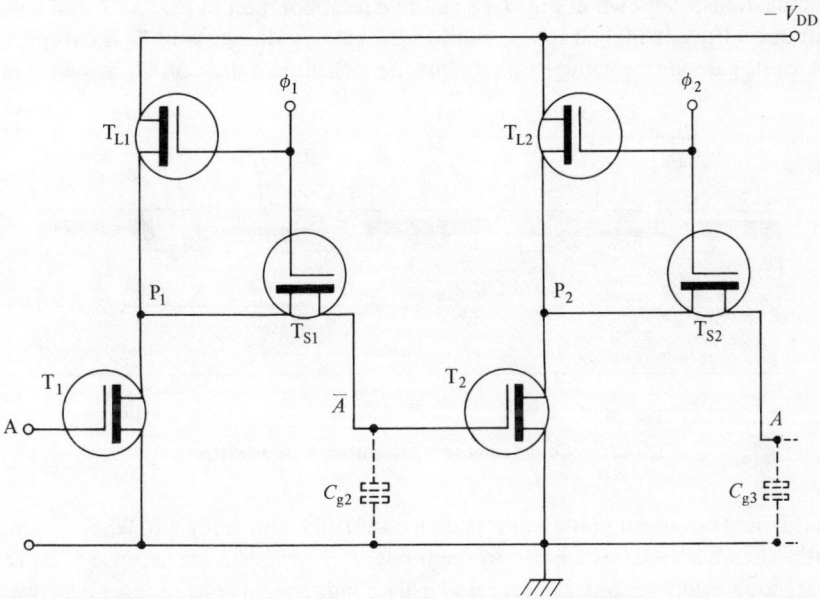

Fig. 7.15   Integrated MOS dynamic shift register.

to power supplies, MOS usually requiring higher d.c. and logic voltages. It is often necessary to interface MOS and bipolar circuits, as, for example, when an MOS shift register is driven by higher speed bipolar logic. As well as the different supply voltage requirements of the two circuits, there is an additional difficulty due to differing logic swings. For example TTL logic might have output voltage swings from 0.5 to 2.5 V, whereas the shift register clock inputs might require switching pulses, say, in the range -10 to -1 V. Circuits to interface TTL to MOS logic have to be included to accommodate these differing requirements.

Alternatively the interfacing problem can be solved by a different integrated technology, for example, the BIGFET circuit described in section 5.10.2.

117

7.10 Complementary MOS logic circuits (COS/MOS)

Logic gates incorporating complementary pairs of MOS transistors, which have one of the lowest power dissipations of any logic family, are increasingly used in large-scale integrated, battery-powered, portable systems such as pocket calculators, or for space applications.

The basic technology for making complementary induced p-channel and n-channel MOSTs on the same integrated chip is illustrated in Fig. 7.16. It will be seen that the p-channel device is self-isolating by virtue of the n-type substrate, but that the n-channel MOST requires an additional p-type isolation diffusion.

The fundamental CMOS logic circuit is an inverter consisting of p-channel and n-channel enhancement mode transistors, connected in series with common drain connections, as shown in Fig. 7.16 and in circuit notation in Fig. 7.17. If a logic 1, around $+V_{DD}$, is applied to the input to the gate, at A, transistor T_1 is cut-off, since its gate-source voltage is zero, but the n-channel transistor, T_2, is switched

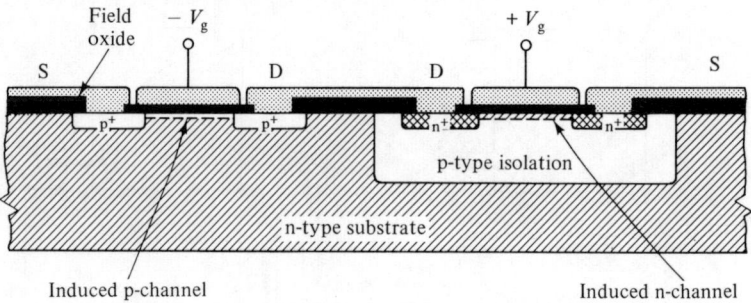

Fig. 7.16   Complementary integrated MOS transistors.

hard on. The output voltage at P is then essentially zero, a logic 0. When, on the other hand, a logic 0 is applied to the input, T_2 is switched off because of its zero gate-source voltage, but T_1 becomes highly conducting due to its large negative gate-source voltage. The output voltage is then around V_{DD}, a logic 1.

Since one or other device is switched off at any part of the logic cycle, only leakage currents flow, and the current drained from the supply is always very small. This leads to a very low power dissipation, typically of order 10 nW in either logic state. A further advantage is that CMOS gates operate from positive voltage supply rails, which is convenient for interfacing with bipolar integrated circuits. If required, however, it is possible to reverse the positions of the two component transistors of the complementary pair, grounding S_1 rather than S_2; the circuit then operates with negative V_{DD} and logic voltages.

The advantages gained by using complementary circuitry for inverters also applies to derivative logic circuits. Consider, as an example, the CMOS three-input NAND logic gate, shown in Fig. 7.18. If a logic 1, equal to $+V_{DD}$, is applied to all the inputs, A, B, and C simultaneously, the n-channel transistors, T_{4-6}, are all

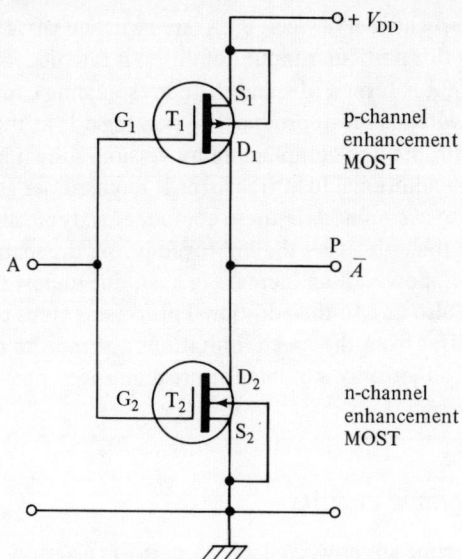

Fig. 7.17   Complementary MOST inverter circuit.

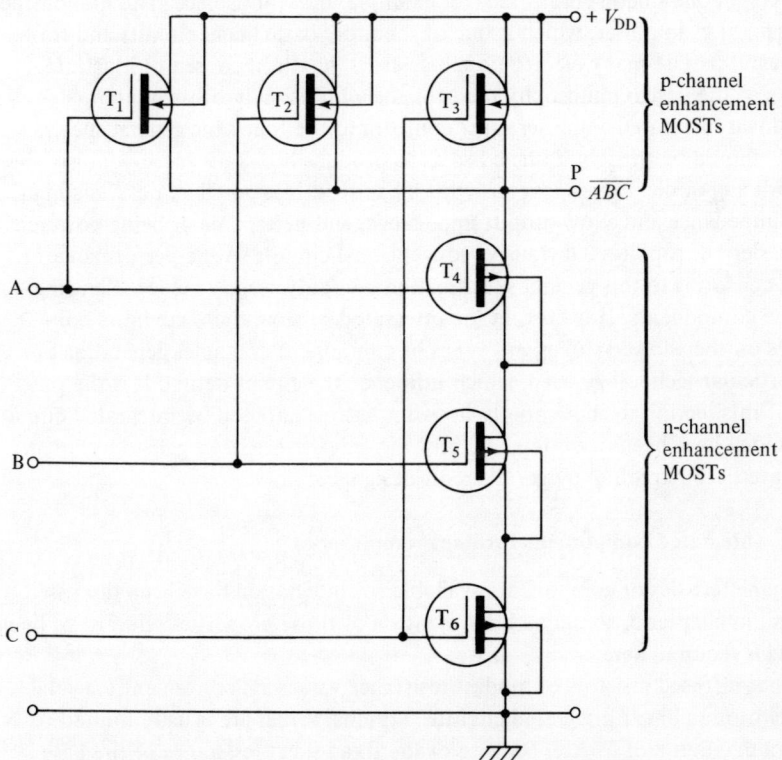

Fig. 7.18   Complementary MOST three-input NAND gate.

119

switched on and the p-channel devices, T_{1-3}, are switched off, so an output logic 0 appears at P. Unless this particular input condition is satisfied, at least one of the series transistors, T_{4-6}, is turned off and the corresponding complementary transistor, T_{1-3}, is switched on to produce V_{DD}, a logic 1, at the output. Comparing the circuit with the non-complementary version, shown in Fig. 7.13(c), demonstrates that an additional load transistor is required per gate. The extra slice area necessary to accommodate these components, typically an increase of 25 per cent, is more than offset by the improved power dissipation for many applications. There is, however, an increase in cost, due in part to the decreased packing density but also due to the additional processing steps required. CMOS logic circuits also suffer from the speed limitations common to other MOS integrated circuits, and are only suitable for medium speed operations, in the range 0–5 MHz.

7.11 Linear integrated circuits

Linear integrated circuits are now available to perform a variety of electronic circuit functions, ranging from simple audio pre-amplifiers to complete radio receivers on one silicon chip — see, for example, the frontispiece. The monolithic, high gain, d.c. amplifier, which is one of the more basic linear circuits and forms an integral part of many electronic sub-systems, has been chosen to illustrate aspects of the design philosophy used for the whole family of such circuits.

A linear integrated amplifier must conform to the following general specification. It should be flexibly designed so as to produce high gain and low noise when fed with a.c. or d.c. signals, over as wide a bandwidth as possible, possess a high input impedance and a low output impedance, and be capable of being powered by a variety of supplies. Obviously, conventional circuits with a performance which conforms to this general specification are fairly readily achievable using discrete components. However, in the integrated version, there are more constraints on the allowed values and types of component available, depending on the particular technology used, which influence the circuit design. It is the purpose of this section to show briefly how limitations imposed by integrated circuit technology have been overcome and how any advantages to be gained by integration have been exploited by good circuit design.

7.11.1 Integrated components for linear amplifiers

The characteristics of components available for integration have been discussed in detail in chapter 5, so only a brief mention of those properties relevant to linear circuits is required here.

Planar diffused resistors of modest resistance values are conveniently available for inclusion in linear integrated circuits. Nominal values are usually limited to below a maximum of 50 kΩ, because of the fixed sheet resistance of the base diffusion in which they are formed and because of the limitations in slice area.

Tolerances of better than ±20 per cent can only be specified at the expense of yield, but ratios of resistances can be more closely controlled. Higher value resistances are usually provided by the dynamic resistance of an active device, as will be discussed later. Whereas low value integrated junction capacitors are available, as described in section 5.1, modern linear circuits are designed so as to avoid their use, partly because of their voltage dependence but mainly to conserve chip area. In the latest generation of designs, an MOS capacitor is included in the circuit chip, as explained later, as part of a frequency compensating network. Because of the limitations discussed in section 5.3, integrated inductors are usually avoided.

Bipolar n–p–n transistors with an acceptably good specification are readily available, using the planar process. A feature of the integration technology is that very close matching of the characteristics of such transistors is possible, and many designs exploit this advantage. Compatible integrated p–n–p devices are available on the same chip, but at the expense of added complication and the possibility of reduced yield, as described in section 5.10, so some of the earlier designs only incorporated n–p–n transistors. More recently designed circuits use complementary transistors in the output stage, and a few lateral p–n–p devices elsewhere in the circuit. However, the p–n–p devices employed are produced by the conventional n–p–n planar process used for the remainder of the circuit, as will be described later.

Unipolar integrated transistors and components are readily available but are not much in evidence in operational amplifier circuits, firstly because of the

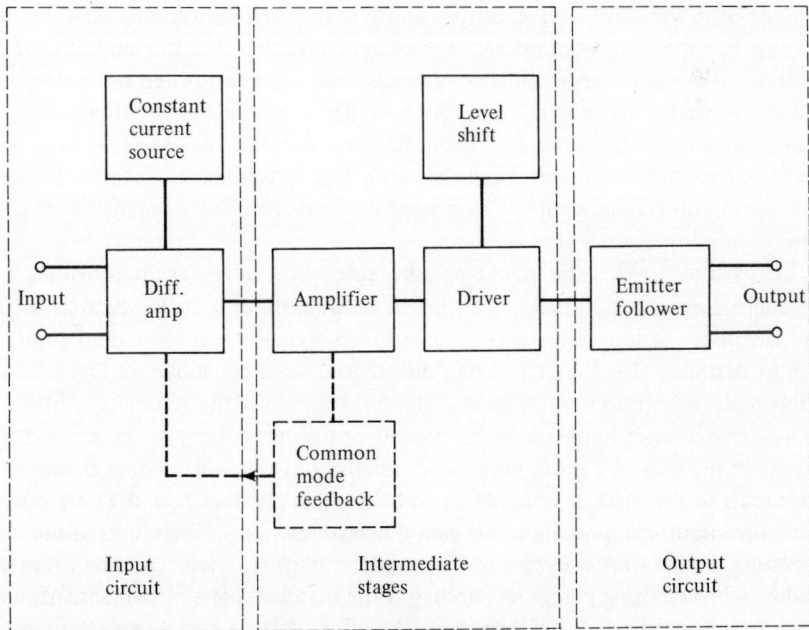

Fig. 7.19 *Basic circuit for an integrated high-gain amplifier.*

121

relatively poor frequency response of all-MOS circuits, and secondly because of the expense of the hybrid technology, described in section 5.10. Integrated JFETs have been incorporated in the input section of some integrated amplifiers to increase their input impedance.

The various integrated diodes available to the designer are discussed in section 5.5. From considerations of convenience and economics, the most common configuration consists of a standard n–p–n transistor with a common collector-base connection, as in Fig. 5.19(d). If Zener diodes are required, these can again be fabricated from the standard planar n–p–n structure, with the emitter-base junction reverse biased, as discussed in section 5.5.3. These integrated components have the disadvantages of a relatively high dynamic resistance compared to the discrete version, only one voltage (~ 6 V) being readily available and a tendency to introduce additional noise. For these reasons, their use is avoided, if possible.

It is evident from these considerations that the design of an integrated linear operational amplifier should preferably be based on bipolar transistors and derivative diodes, plus diffused planar resistors, with a low and not very critically defined resistance.

7.11.2 General circuit description of integrated operational amplifiers

Most available linear integrated circuit amplifiers can be reduced to the standardized form shown in Fig. 7.19, consisting essentially of an input circuit, an intermediate or driver stage, and an output section. This description might also be pertinent to the equivalent discrete component amplifier, but the methods for achieving the required performance for each stage using integrated technology and circuit design are usually quite different. The d.c. coupling requirement is readily achieved in the integrated amplifier, thus avoiding the use of integrated coupling capacitors and conserving chip area. It is convenient to examine briefly how the circuit requirements of each stage might be fulfilled using planar monolithic technology.

Let us consider first the input stage. Because of the advantage in providing a differential input circuit and in view of the desirability of reducing thermal drifts and sensitivity to supply variations, some sort of symmetrical input configuration is most desirable. The long-tailed pair differential amplifier shown in Fig. 7.20 offers many advantages and is most common. Because of the symmetry of the circuit, any changes in the circuit to the left of the centre line, for example, a change in the gain of T_1, will tend to be equalized by a corresponding change in the circuit to the right. If resistors R_1 and R_2 and transistors T_1 and T_2 are matched pairs, the circuit can provide stable gain down to low frequencies, even if the operating temperature is cycled or the supply voltage is varied. This condition is readily achieved using planar technology if the matched pairs of components are sited close together on the silicon chip. The offset voltage, which is the voltage at the input terminals of a differential amplifier that is necessary to reduce the

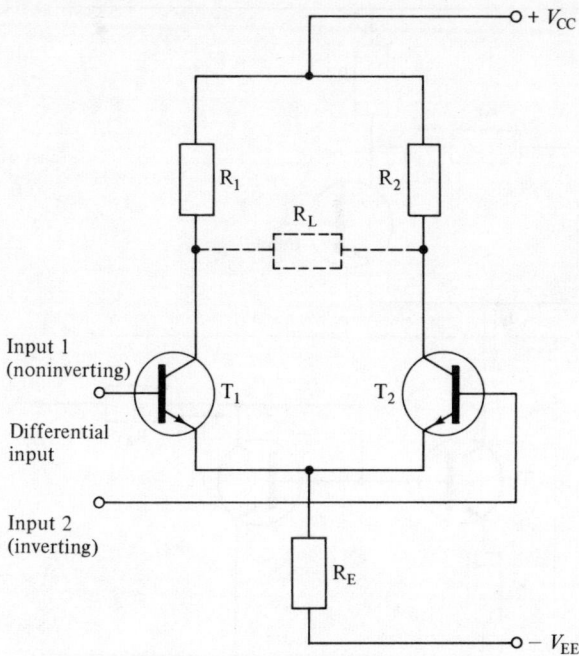

Fig. 7.20   Basic differential input circuit.

output voltage to zero, is also much reduced by successful matching of components.

In order to obtain good common mode rejection, the emitter resistor, R_E, should have a high value. This condition is not easily satisfied using a diffused integrated resistor, so, in the integrated circuit, a low-level constant current source replaces R_E. A possible typical circuit arrangement is shown in Fig. 7.21. In this circuit, the correct choice of R_3 can result in a very small constant current, I_1, in the micro-amp range, for relatively small values of R_3, of only a few kilo-ohms. If common mode feedback is required, this may be brought from a suitable point in the intermediate stage to point A.

In order to achieve a higher input impedance, the differential transistors, T_1 and T_2, are sometimes each replaced by a Darlington pair, shown in Fig. 7.22. There are, however, some disadvantages to this input arrangement, such as increased thermal sensitivity and offset, which have to be considered. Notice that the two transistors have common collectors, so in integrated form can share the same collector diffusion.

Some of the functions which may be performed by the intermediate stages of an integrated operational amplifier are (a) to provide further amplification, (b) to convert a differential input connection to the single-ended signal required at the input of the output stage, thus saving chip area, (c) to provide a common mode feedback point, and (d) to provide any level shifting necessitated by the input

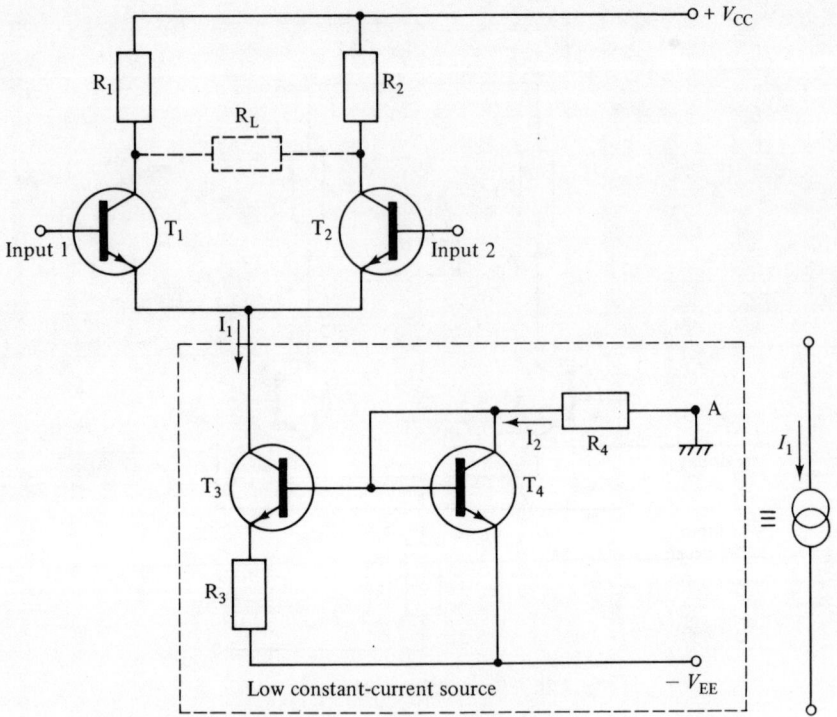

Fig. 7.21 Integrated differential input with constant current circuit.

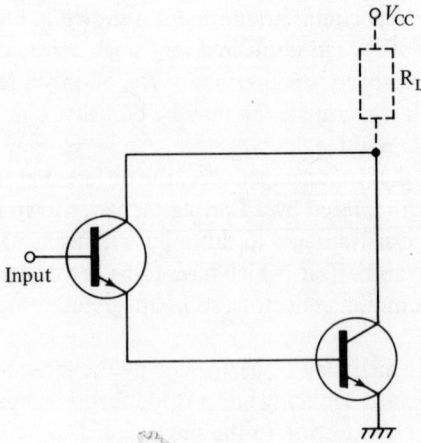

Fig. 7.22 The Darlington connection.

124

requirements of the output stage. A possible simplified circuit configuration of a second stage which performs functions (a) and (b) above is shown in Fig. 7.23. Transistor T_4 provides second-stage amplification and a single-ended output. Transistor T_3 is a unity gain amplifier which inverts the output of T_1 and combines it with the output of T_2, so that the full differential gain of the input stage is employed. In a more realistic circuit transistors T_3 and T_4 in Fig. 7.23 are sometimes replaced by pairs of transistors in a modified Darlington connection, to provide additional current gain and prevent loading of the input circuit. Close matching of the parameters of the Darlington pairs and their nominally identical load resistors, R_3 and R_4, is again achieved by placing these components as close

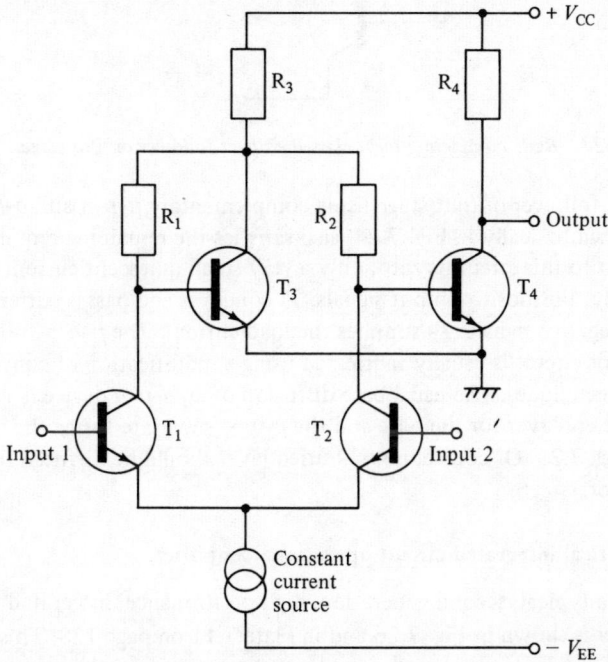

Fig. 7.23   *Possible input and second stages of an integrated amplifier.*

together as possible on the integrated circuit chip. The output from the second stage is normally required to be at a quiescent level midway between two supply lines, providing positive and negative output swings. This condition is clearly not fulfilled by the simplified circuit of Fig. 7.23 and additional level shifting circuitry becomes necessary.

The required general properties of the output stage of an integrated operational amplifier are (a) a low output impedance, (b) a low quiescent power, (c) zero quiescent output voltage, and (d) a large available output voltage swing, approaching the supply line voltages, with the minimum of distortion. The low output impedance requirement is best met by some sort of emitter follower circuit. The

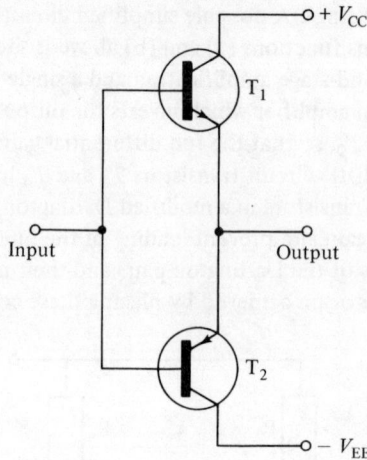

Fig. 7.24 Basic complementary, class-B emitter-follower output stage.

class B emitter-follower output stage using complementary n-p-n and p-n-p
transistors shown basically in Fig. 7.24, also satisfies the remaining requirements.
When the input to this circuit is zero, only a very small quiescent current is drawn
from the supply. For positive input signals, T_1 conducts and passes current to the
load and for negative signals, T_2 supplies the load current. The p-n-p transistor
required for this circuit is usually fabricated using a modification of conventional
n-p-n planar techniques. The usual base diffusion of an n-p-n device is used for
an emitter, the epi-layer for the base and the p-type substrate forms the collector,
as shown in Fig. 7.25. Of course, no n^+ buried layer should be diffused for this
p-n-p transistor.

7.11.3 A practical integrated circuit operational amplifier

The circuit of a typical, second generation, high performance, integrated operational amplifier is shown in Fig. 7.26 and in Plate 7.11 on page 128. This circuit

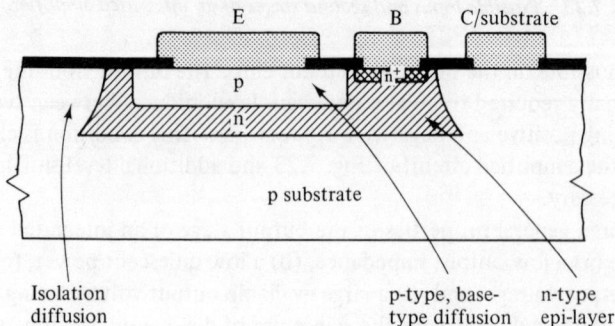

Fig. 7.25 An integrated p-n-p transistor used for a complementary output stage.

126

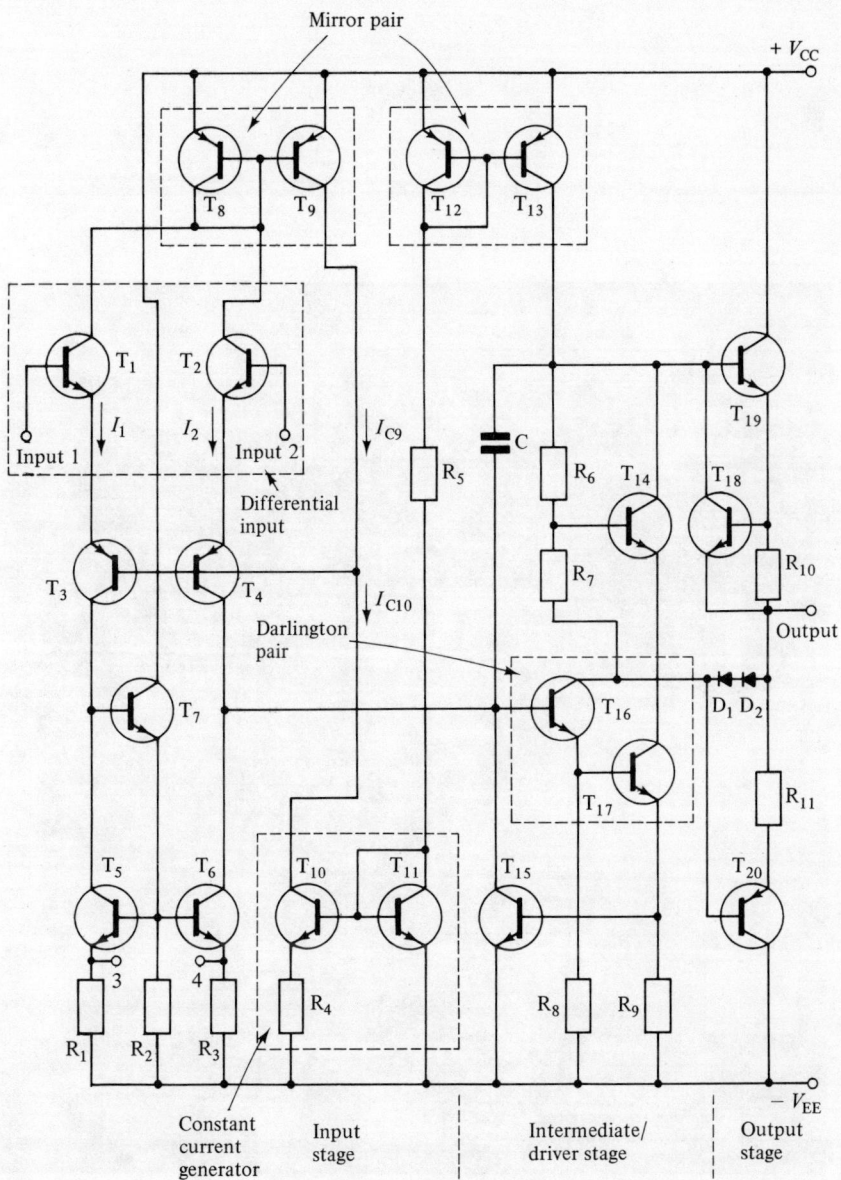

Fig. 7.26 Integrated high-gain amplifier circuit.

127

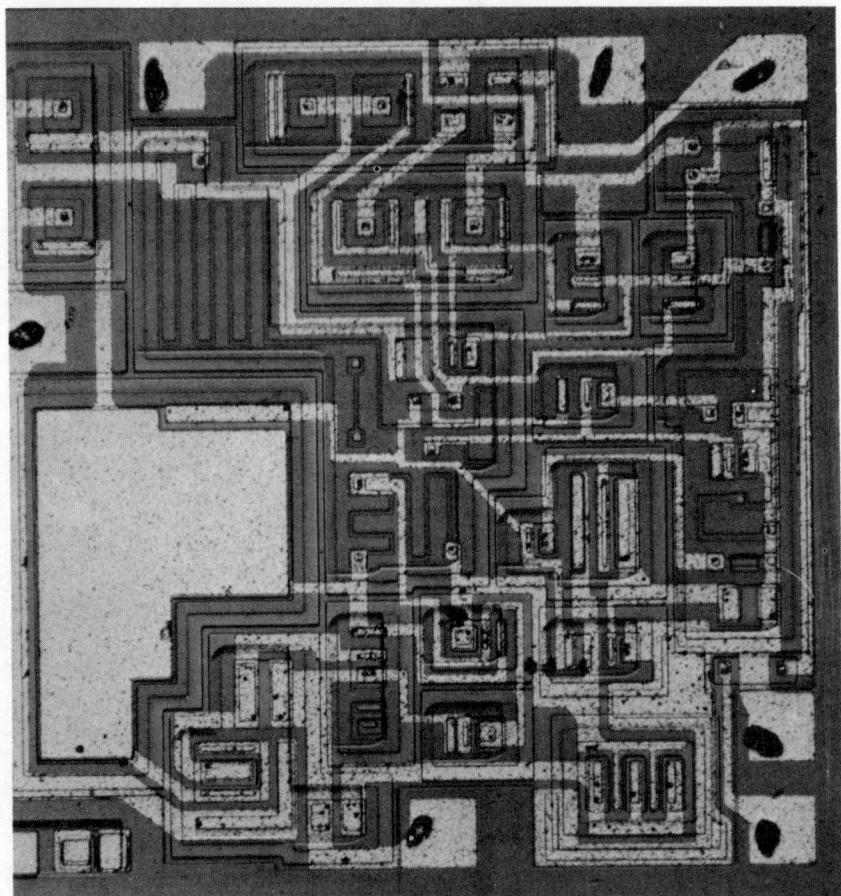

Plate 7.11   A microphotograph of a 741 monolithic integrated operational amplifier chip. Its equivalent circuit is shown in Fig. 7.26. Note the large area MOS capacitor. Reproduced by kind permission of GEC Semiconductors Ltd.

128

embodies many of the design concepts outlined in the previous section. While a detailed circuit analysis is out of place here, some of the additional circuit and technological features are relevant.

It will be seen that the circuit can be essentially subdivided into a differential input stage, with principal components T_1 and T_2, which feeds an intermediate amplifying stage, consisting of a Darlington pair, T_{16} and T_{17}, which in turn drive a complementary emitter-follower output stage, T_{19} and T_{20}.

The detailed circuit design includes several *current mirrors*, which consist of matched transistor pairs, connected as illustrated in the example shown in Fig. 7.27. It will be noted that the two transistors have identical emitter-base voltages, so have equal collector currents; one collector current always mirrors the other. For such a circuit to operate efficiently, the constituent transistor pair are required to be as nearly identical as possible, thus ensuring that they have the same collector current-base voltage relationship. The circuit is therefore particu-

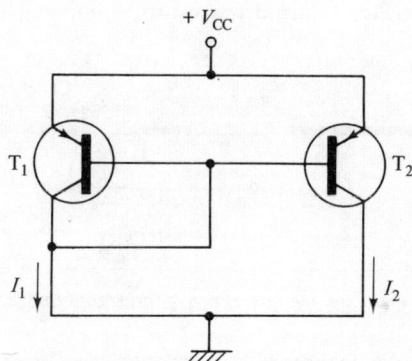

Fig. 7.27   An example of the current mirror configuration. $I_1 = I_2$.

larly well suited to the monolithic planar process, particularly if the matched transistors are located close to each other.

Returning to the complete circuit, it will be noticed that transistors T_8 and T_9 constitute a current mirror. Thus the collector current of T_8, which supplies the input transistors T_1 and T_2, is identical to, or mirrored by, the collector current of T_9. Or, using the notation in Fig. 7.26 $I_{C9} = I_1 + I_2$. Further, the mirror pair, T_{10} and T_{11} in conjunction with R_4 form a low constant current source supplying I_{C10}. This current is almost equal to I_{C9}, since the base currents drawn by T_3 and T_4 are negligible. By this means, the low constant current source effectively supplies the total current to the input pair, $I_1 + I_2$.

Transistors T_{12} and T_{13} are also connected in a current mirror configuration, so that T_{13} behaves as a very high load impedance to the Darlington pair output driver stage.

Transistors T_5 and T_6 behave as high impedance loads for the input transistors, having effective values of order 1 MΩ, and so ensure high differential voltage gain from this stage. Such high values of resistance would not be feasible using diffused planar resistors.

Another feature of the circuit is the use of low h_{fe} p-n-p transistors, for example T_3 and T_4, which are used here for level setting. We have seen that although the production of high h_{fe} complementary p-n-p devices on the same slice as n-p-n devices is possible, it automatically involves costly additional processing steps. Wherever necessary then, *lateral p-n-p* transistors constructed as shown in Fig. 7.28, are incorporated in the circuit. Such transistors are formed, as shown, using conventional n-p-n planar technology, but the active transistor currents flow laterally, parallel to the surface of the chip. The very wide effective base width, which corresponds to a part of the usual collector region, accounts for the degraded performance of the p-n-p device. Transistors T_3 and T_4 are connected in a common base configuration, so as to improve their limited frequency response.

Transistor T_{14} and its associated circuitry maintains a low, constant quiescent current flowing through the output transistors, so minimizing cross-over distortion.

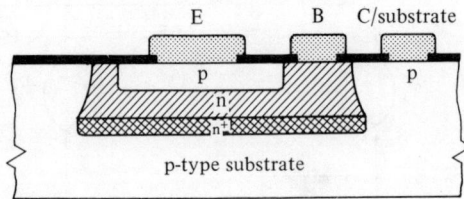

Fig. 7.28   A lateral p-n-p integrated transistor.

The output stage also incorporates current limiting circuitry to prevent excessive dissipation, in the event of the output transistors being short-circuited. If the output current flowing through T_{19}, which is monitored by resistor R_{10}, becomes equal to a design maximum, transistor T_{18} is turned on, diverts base current from T_{19} and so turns it off. Diodes D_1 and D_2, which in practice are transistors with a common collector-base connection, are used in conjunction with R_{11}, to provide similar protection for the output transistor T_{20}.

Capacitor C is included in the circuit to provide frequency compensation. Although its capacitance is only of order 30 pF, its effective value is very much higher, due to the relatively high impedance level occurring at the output of T_4 and the Miller action of the driver stage. This effective capacitance is sufficient to produce a break point at approximately 10 Hz and a 20 dB per octave roll-off down to unity gain, thus ensuring that the amplifier can be used with 100 per cent negative feedback, for example in a voltage follower circuit, without the possibility of instability. In earlier integrated circuits, it was necessary to add capacitance externally, but it has been found possible in this more sophisticated version to include an integrated MOS capacitor fabricated on the same chip as the amplifier. The technology used to produce such a capacitor, which is illustrated, for example, in Fig. 5.1, is entirely compatible with the conventional planar techniques used to produce the rest of the circuit, provided ultra-clean production techniques are employed.

7.12 Charge-coupled device circuits (CCD)

There follows a brief description of the construction and operation of a charge-coupled device, as an example of a complete integrated circuit that performs a unique circuit function which cannot be exactly reproduced by any circuit employing discrete components. The circuit, or device, is shown in part in its most elemental form in Fig. 7.29. It will be seen that the structure consists essentially of a series of metal gate electrodes, separated from an n-type semiconducting substrate (for a p-channel device) by a thin oxide layer. It will be evident that such structures are in principle fairly straightforward to make using conventional MOS technology as discussed in section 5.6.2, but there are some additional difficulties that will be mentioned later.

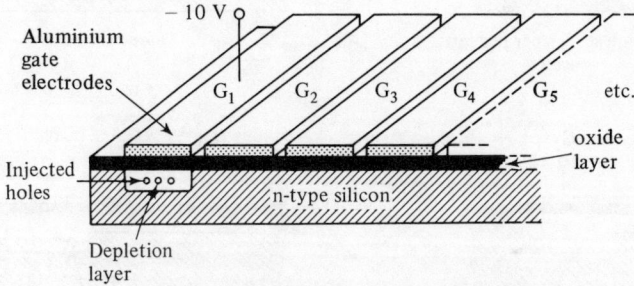

Fig. 7.29   The basic charge-coupled device structure.

If a negative-going pulse, of magnitude say -10 V, is applied to the first gate, G_1, a depletion layer is initially formed, in typically less than 1 μs, as shown. Since the gate voltage is greater than the turn-on voltage, the additional charge required at the oxide–silicon interface would eventually be provided by minority holes which move into the region to form the usual inversion layer. This process takes a relatively long time, typically of the order 1 s, and is usually inhibited. In times short compared with one second, say in the first few milliseconds after the application of the pulse, a non-equilibrium situation exists, in which the inversion layer has not had time to form, but a depletion layer exists, as shown in Fig. 7.29. This layer behaves as a potential well and if minority holes are injected into it in some way, as shown, these will become trapped on the well. The trapped holes can then be transferred to a well under G_2, by a process to be described, and then to under G_3 and so on, sequentially down the whole structure. The device thus performs as a dynamic shift register. This process is only possible provided the charge is not stored in any well long enough for an inversion layer to form, in which case the charge would disappear and the corresponding information would be lost. The

131

CCD structure thus behaves as a dynamic shift register, and charge has to be transferred in times short compared to one millisecond, say.

A three-phase clocked voltage pulse system supplied to the gates ensures that charge is transferred serially between gates and its direction is controlled, as follows. The first phase connects a -10 V pulse to G_1, to produce the potential well into which information, in the form of minority holes, is stored, as described. During phase two, the adjacent gate, G_2, is biased to a greater negative voltage, say -15 V, to produce a deeper well under it, as shown in Fig. 7.30(a). The stored charge then transfers into the deeper potential well by diffusion down the potential gradient, which incidentally can be a relatively slow process. Note that, to ensure charge transfer, the potential wells must physically overlap. As depletion

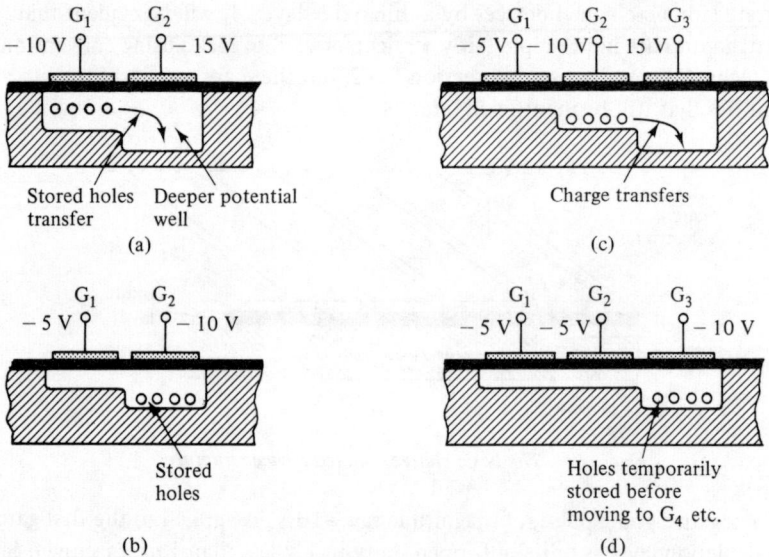

Fig. 7.30   *The mechanism of charge transfer in a three-phase CCD.*

layers are typically only a few micrometres deep, the spacing between neighbouring gates has to be as small as possible, to ensure adequate overlap. The charge is then completely stored in the well under G_2, so the voltage on G_1 is reduced to a low value, say -5 V, and that on G_2 to a sustaining level of, say, -10 V, as shown in Fig. 7.30(b).

A third phase transfers a -15 V voltage pulse to the next gate, G_3, and the charge is transferred from G_2 to the well under it, as shown in Fig. 7.30(c). The voltages on G_2 and G_3 can then be relaxed as before, as shown in Fig. 7.30(d) to complete one cycle of the clock frequency. The charge has been transferred from under G_1 to under G_3 in one cycle of the clocked three-phase pulses, which causes a series of voltages in the sequence of $-15, -10, -5, -15$, etc., to be applied to each gate electrode. The charge can be transferred in this manner down the structure, each storage cell of three adjacent electrodes accommodating the *bit* of information. As soon as charge is moved out of one set of three electrodes, say from G_3 to G_4, then the input gate is again put in a state to receive a further *bit* of information.

The practical realization of an 8-bit CCD shift-register is shown diagrammatically in Fig. 7.31. The aluminium gate electrodes are typically 250 x 50 μm and the gaps between them 2 μm, which in itself is difficult to achieve technologically. Since there are three gate electrodes per storage cell, every third one is connected in parallel and connected to one phase of the three-phase supply, ϕ_1, ϕ_2, ϕ_3, as shown. Because of topological difficulties, one set of connections to a common phase is often by means of a diffused cross-under, as described in section 6.2, but the remainder of the connections can be done by the usual metallizing process. The gate oxide is typically only 100 nm thick but the field oxide elsewhere is about 500 nm thick.

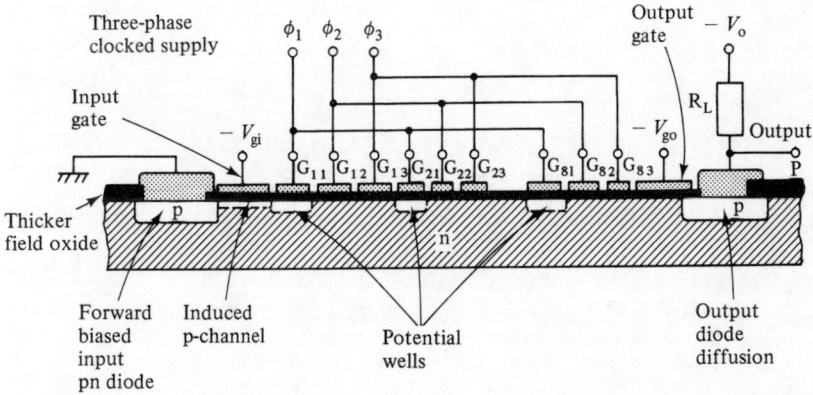

Fig. 7.31   An 8-bit CCD dynamic shift register.

Input holes, generated by the forward biased diffused input diode, are injected into the potential well under the first gate G_{11}, via an induced channel under the input gate, if this is biased sufficiently negative, as shown in Fig. 7.31. In other words, the potential well under G_{11} behaves as an equivalent drain in a MOST that is controlled by the input gate. The injected charge is then transferred down the structure by the clocked three-phase transport mechanism described and its presence at the output is detected by the reverse biased diffused output diode shown. The minority holes that arrive under electrode G_{83} are switched to the output diode via the induced channel under the output gate. They are then swept across the reverse biased *p–n* junction to produce a corresponding current in the load resistor, R_L, and a signal output.

The input, output, and information shifting schemes described are only typical of several alternative arrangements. For example, it is possible to dispense with the input diode and to introduce minority carriers under G_{11}, say, by irradiation of

133

the silicon by light, which produces holes there. The charge that is temporarily stored and is subsequently transferred down the structure produces an output which is proportional to the intensity of the illumination. A matrix version of a CCD, operated in this mode, has considerable potential in the imaging field and a solid-state colour TV camera, which uses the same principle, has already been demonstrated.

Another possible application of CCDs, which is receiving considerable attention, is concerned with the processing of analogue signals. It is fairly straightforward to envisage the input of a simple linear CCD array sampling a slowly varying analogue signal, applied to its input gate, which can be reconstituted at the output of the CCD, after having experienced some delay. The delay can be varied over a wide range, electronically, by varying the clock frequency. It is also possible to process an analogue signal further, as it progresses along a CCD structure, by, say, tailoring the geometry of the phase gates.

Bibliography

Suggestions for further reading:

Allison, J., *Electronic Engineering Materials and Devices*, McGraw-Hill.
Camenzind, H. R., *Circuit Design for Integrated Electronics*, Addison Wesley.
Fitchen, F. C., *Electronic Integrated Circuits and Systems*, Van Nostrand Reinhold.
Ghandi, S. K., *The Theory and Practice of Microelectronics*, Wiley.
Khambata, A. J., *Large Scale Integration*, Wiley.
Penney, W. M. and Lau, L., *MOS Integrated Circuits*, Van Nostrand.
Runyon, W. R., *Silicon Semiconductor Technology*, McGraw-Hill.
Warner, R. M. and Fordemwalt, J. N., eds. *Integrated Circuits—Design Principles and Fabrication*, McGraw-Hill.

Index

Printed by William Clowes & Sons Limited, London, Colchester and Beccles